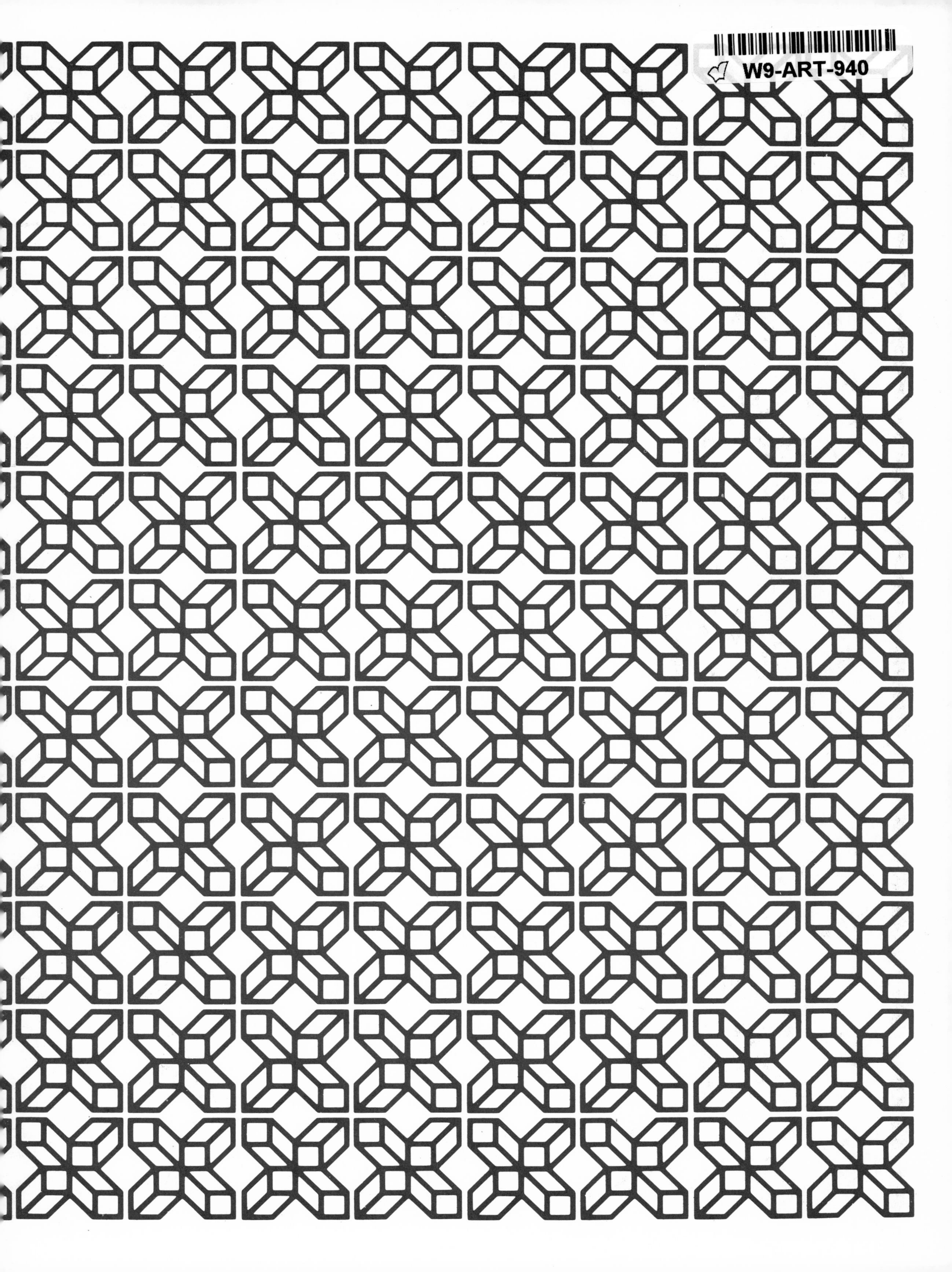
W9-ART-940

Perspectives in Geography 3
THE NATURE OF CHANGE IN GEOGRAPHICAL IDEAS

NORTHERN ILLINOIS UNIVERSITY

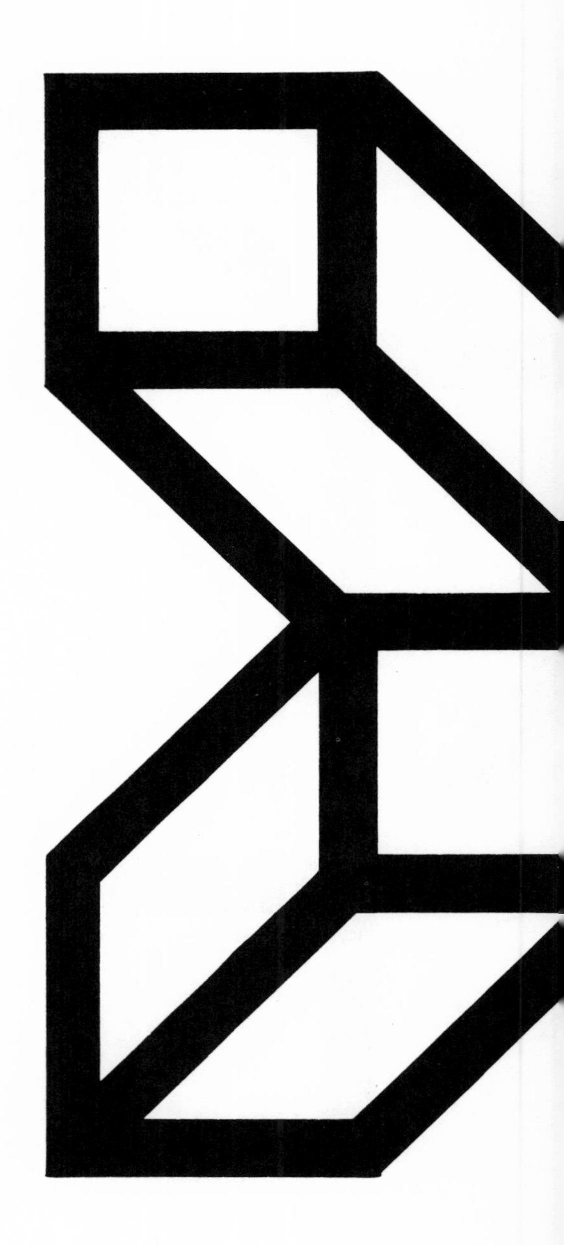

NORTHERN ILLINOIS UNIVERSITY PRESS

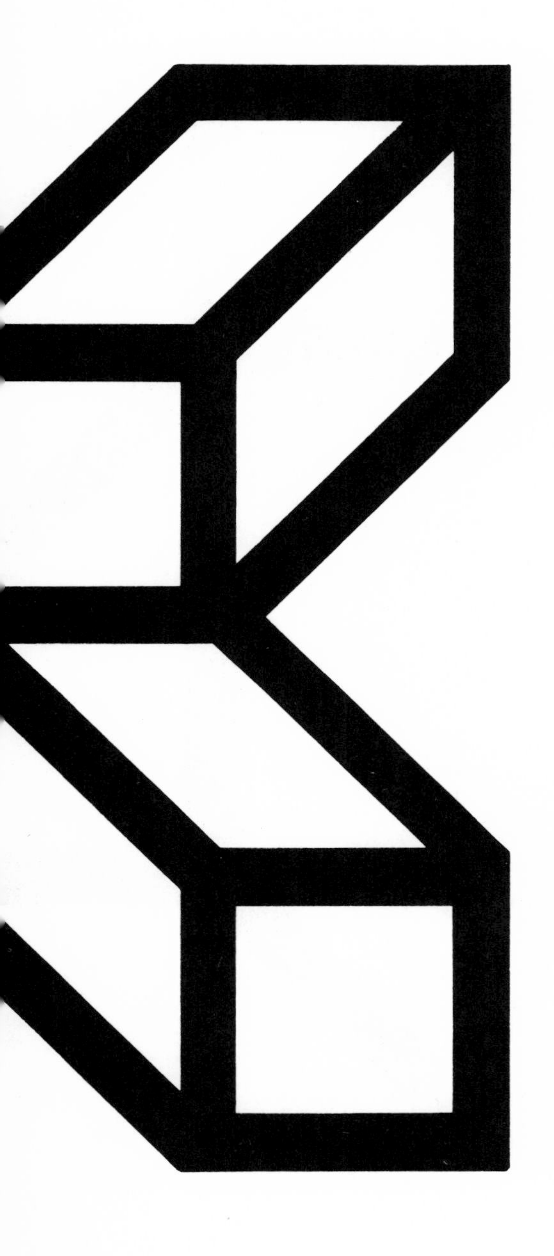

Perspectives in Geography 3

THE NATURE OF CHANGE IN GEOGRAPHICAL IDEAS

VOLUME EDITOR

BRIAN J. L. BERRY

GENERAL EDITOR

BRIAN J. L. BERRY

Brian J. L. Berry is Williams Professor of City and Regional Planning and Director of the Laboratory for Computer Graphics and Spatial Analysis at Harvard University.

CONTRIBUTING AUTHORS

Brian J. L. Berry, Harvard University; Donald W. Jones, University of Chicago; Marvin W. Mikesell, University of Chicago; Christopher F. Müller-Wille, University of Chicago; Thomas R. Tocalis, University of Chicago.

Marvin W. Mikesell's "The Rise and Decline of Sequent Occupance" is from *Geographies of the Mind: Essays in Historical Geosophy,* edited by David Lowenthal and Martyn J. Bowden. Copyright © 1975 by Oxford University Press, Inc. Reprinted with permission.

Library of Congress Cataloging in Publication Data
Main entry under title:
The nature of change in geographical ideas.
(Perspectives in geography; 3)
1. Geography–Philosophy–Addresses, essays, lectures. I. Berry, Brian Joe Lobley, 1934-II. Series.
G70.N37 910'.01 75-39294
ISBN 0-87580-063-7

Published by the Northern Illinois University Press,
DeKalb, Illinois 60115
Manufactured in the United States of America

TABLE OF CONTENTS

INTRODUCTION: A KUHNIAN PERSPECTIVE — vii
Brian J. L. Berry, Harvard University

1
THE RISE AND DECLINE OF SEQUENT OCCUPANCE — 1
Marvin W. Mikesell, University of Chicago

2
GEOGRAPHICAL THEORIES OF SOCIAL CHANGE — 17
Brian J. L. Berry, Harvard University

3
THE FORGOTTEN HERITAGE: CHRISTALLER'S ANTECEDENTS — 37
Christopher F. Müller-Wille, University of Chicago

4
CHANGING THEORETICAL FOUNDATIONS OF THE GRAVITY CONCEPT OF HUMAN INTERACTION — 65
Thomas R. Tocalis, University of Chicago

5
IMPLICATIONS OF "SCHOOLING" IN ECONOMIC ANTHROPOLOGY FOR INTERPRETATIONS OF THE ECONOMIC GEOGRAPHY OF NONINDUSTRIAL SOCIETIES — 125
Donald Jones, University of Chicago

REFERENCES CITED — 154

INTRODUCTION: A KUHNIAN PERSPECTIVE

BRIAN J. L. BERRY

What has been the nature of change in geographical ideas? Immediately following World War II, such a question would have been answered with many references to Richard Hartshorne's study, *The Nature of Geography.* Progress in geography, in Hartshorne's terms, involved completion of more comprehensive and up-to-date empirical studies conducted according to the tenets of a mainstream philosophy of areal differentiation. Yet, by the mid-1960s, the situation had changed drastically. The Hartshornian mainstream had withered under theoretical-quantitative attack. To a new generation of young geographers, progress in geography meant theory-building, theory-testing, and theory-refining in an accumulative and unidirectional sequence, and the theory was location theory—von Thünen, Weber, Christaller, and Lösch.

Had the philosophy of areal differentiation been tested and found wanting, or did something else happen to change the focus, content, and orientation of geographic research? Was the change in geographical ideas that took place different from the disciplinary reorientation that accompanied the replacement of environmental determinism as a central concept by areal differentiation in the 1920s?

Until the publication, in 1962, of Thomas S. Kuhn's *The Structure of Scientific Revolutions* (Chicago: University of Chicago Press, International Encyclopedia of Unified Science) we would have been hard pressed to answer these questions, for Kuhn's work marks a significant watershed in our understanding of the nature of scientific progress. Seen from Kuhn's perspective, the rise of areal differentiation involved the abandonment of science for a while by geographers, as one paradigm, environmental determinism, fell into disfavor. Then, buffeted by a quantitative revolution led by young spatial analysts, the recognition of new sources of geographical theory led the discipline to its present state. Approval and disapproval, favor and disfavor are the key themes illustrated in the essays in this collection. The changes in geographical ideas that we have discussed are distinctly Kuhnian.

I. KUHN ON THE NATURE OF SCIENTIFIC PROGRESS

Kuhn's genius was that he saw scientific progress not in cumulative or unidirectional terms but as successive periods of "normal" science interspersed by revolutionary paradigm shifts. Normal science arises when a scientific community accepts that one or more achievements of the past supply the foundation for its further practice. Mutual acceptance of a common framework sets the stage for theory-building and theory-testing as separate activities, because both theory and fact are determined by the nature of that framework.

Such consensus is not easily achieved. In the early stages of the development of any science, different men confronting the same range of phenomena (although not usually all the same particular phenomena) describe and interpret them in different ways, creating different, competing schools of thought. What Kuhn found was that in most scientific fields such divergences ultimately disappear, usually because of the triumph of one of the schools which, like every other school, has its own characteristic beliefs and preconceptions and emphasizes only some special part of the sizable and inchoate pool of information that is the "real world." To be accepted as a paradigm, a particular school's theory, Kuhn argued, must seem better than its competitors, but it need not, and in fact never does, explain all the facts with which it can be confronted.

When a synthesis is produced that attracts most of the next generation's practitioners, the older schools gradually disappear. Specialized journals and societies appear, and there are claims for a special place in the curriculum, as when the present institutional pattern of scientific specialization developed at the end of the nineteenth century. There are many consequences. When a scientist takes a paradigm for granted he need no longer attempt to build his field anew, starting from first principles and justifying each concept used—such is the task of textbooks. Instead he can practice normal science, devoted to the articulation of the phenomena and theories already supplied by the paradigm by: (1) measuring the facts deemed to be important to the paradigm more precisely; (2) comparing the facts with the predictions made, using the paradigm theory; and (3) conducting additional empirical work to articulate the paradigm theory, resolve its residual ambiguities and solve problems to which the theory had previously only drawn attention. As Kuhn notes, these three classes of problems—determination of significant facts, matching of facts with theory, and articulation of theory—exhaust the literature of normal science, both empirical and theoretical.

For there to be normal science, there must be an accepted paradigm. How, then, does a paradigm achieve priority? The preparadigm period provides keys, marked as it is by frequent debates over legitimate methods, problems, and standards of solution—debates that reappear just before and during scientific revolutions. Paradigms come under attack and are subject to change, according to Kuhn, where the accepted theories prove incapable of explaining existing facts, or because new theories provide an alternative world view. For while the acceptance of a particular paradigm provides all of the benefits of professionalism—efficiency, accuracy, and the like—it also produces highly focused tunnel vision and extreme conservatism with respect to new or alternative ideas. Proponents of alternative paths are thus dismissed as malcontents, kooks, or worse—witness Richard Hartshorne's dismissal of competing views in *The Nature of Geography* and *Perspective on the Nature of Geography.*

Yet persistent anomaly or insistent revolutionaries can progressively erode attachments to an orthodoxy, especially those by the young; and in the crisis conditions that ensue the alternate theories vie for ascendancy, as large-scale paradigm destruction occurs and major shifts take place in the acknowledged nature of the legitimate problems and techniques of normal science. When the paradigm shift finally occurs, it comes not because existing scientists have tested the alternate theories, found the old wanting, and thus disproving, have turned to the new; it comes because the old have been disapproved and simply gone out of style. A paradigm functions, says Kuhn, by telling the scientist about the entities that nature does and does not contain and about the ways in which those entities behave. That information provides a map whose details are elucidated by normal scientific research. And since nature is too complex and varied to be explored at random, that map is as essential as observation and experiment to science's continuing development. Through the theories they embody, paradigms determine the nature of research activity, providing scientists not only with a map but also with some of the directions essential for mapmaking. In learning a paradigm the scientist acquires theory, methods, and standards together, usually in an inextricable mixture. Therefore, when paradigms change, there are usually significant shifts in the criteria determining the legitimacy both of problems and of proposed solutions.

The choice between competing paradigms involves questions that cannot be resolved by the criteria of normal science; adherents of competing schools find it impossible to define common terms facilitating rational debate. Questions of values arise that have to be answered outside normal science altogether. In short, a paradigm shift involves a change in world view. Led by a new paradigm, scholars adopt new methods, look in new places, and even see new and different things when looking with familiar methods in places they have looked before. There is a gestalt switch that takes place all together, or not at all.

II. PROGRESS IN GEOGRAPHY

What, then, is progress in geography? The perspective provided by the essays in this book is distinctly Kuhnian. Marvin W. Mikesell tells the story of the rise and decline of sequent occupance, at its simplest only an elementary organizational concept, but, appearing during the period of disapproval of environmental determinism, one that offered to a significant school of geographers an alternative theory centering on historical determinism. That this school did not prevail was not because an alternative scientific paradigm replaced it, but because, in rejecting the unilinear evolutionism of the social Darwinists, geographers rejected explanation itself. The orthodoxy of areal differentiation advocated research in the absence of guiding theory, but as Kuhn himself notes, "to reject one paradigm without simultaneously substituting another is to reject science itself." Sequent occupance studies degenerated into successive descriptive snapshots and, as Mikesell says, the idea "evolved from novelty to relic without ever being subjected to critical review."

In my essay I also subscribe to the Kuhnian view, as I tell the story of the emergence of geography's first paradigm, *the* geographical theory of environmental determinism that was so central to the emergence of geography as a professional field at the end of the nineteenth century, and relate the conditions which ultimately led to its replacement by another geographic paradigm of social change.

In the same vein, Müller-Wille examines the antecedents of Christaller, whose seminal contributions were so central to geography's recent theoretical-quantitative revolution. Müller-Wille outlines the emergence of urban geography in Germany, the rejection of the Ratzelian unitary interpretation of the field, the outlining by Schlüter of a model consistent with the Hettnerian orthodoxy, the emergence of a series of major findings relating to problems then considered to be marginal to urban geography, and the codification of these findings by Christaller in his new, abstract, deductive theories of settlement size and location.

Few such changes have occurred in geography in isolation from the rest of science, and indeed many geographical concepts have proven to have great resiliency because of their mutability, as the prevailing philosophy of science has changed, finding successive reinterpretation in the changing frameworks of science generally. So it is with the gravity concept of human interaction, the concern of Tocalis's essay. Tocalis relates the course of the model from its initial conception in the minds of Greek philosophers through many manifestations to its present treatment as an extension of statistical mechanics and of general relativity physics.

Borrowing from other sciences is not without its difficulties, even if it does open up opportunities. In the concluding essay Jones explores the problems that arise as cultural geographers interpret the economic geography of nonindustrial societies, using concepts borrowed from the substantivist school of economic anthropology. The purpose of his essay is to propose an interpretive shift, by bringing to bear the concepts and tools of modern economics. In so doing, he exposes what is central to such a shift, and indeed to any understanding of the nature of progress in geography—the priority that must be accorded to changes in world view as opposed to the unimaginative nit-picking that so frequently characterizes normal scientists practicing normal science.

1

THE RISE AND DECLINE OF SEQUENT OCCUPANCE

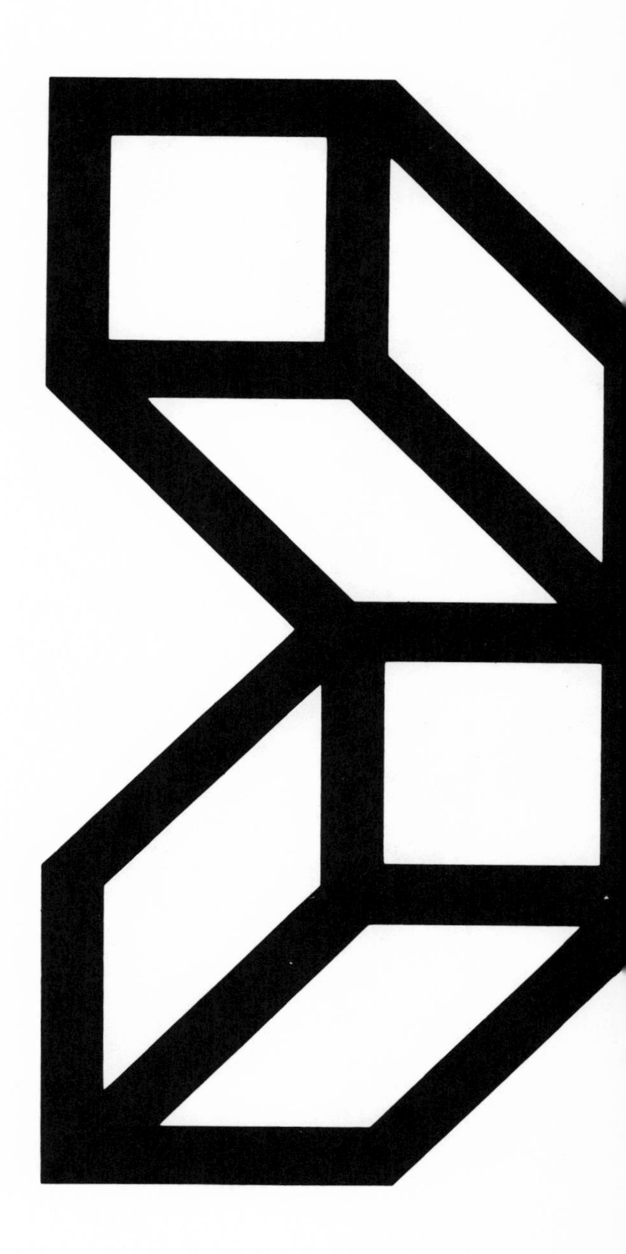

Marvin W. Mikesell
University of Chicago

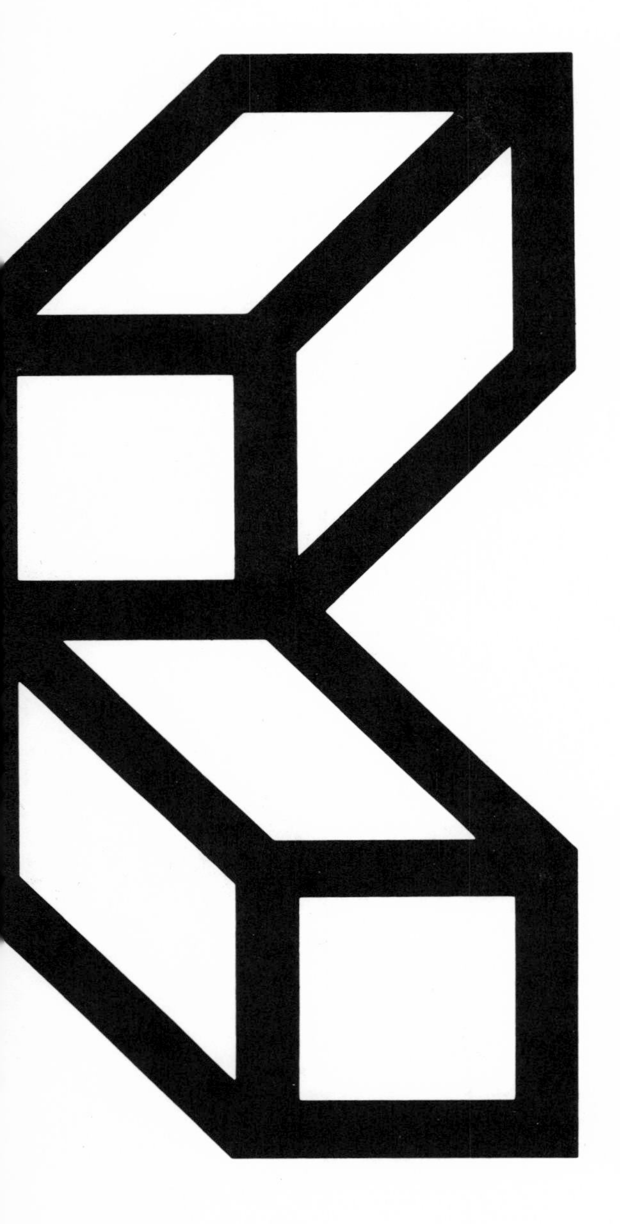

THE RISE AND DECLINE OF SEQUENT OCCUPANCE

> Among the most valuable lessons to be learned from the history of science are those concerning the ways in which science has hitherto reflected human nature and will doubtless continue to reflect it. John K. Wright (1944)

Historians of science usually display confidence in their findings; for the new in their inquiries inevitably seems superior to the old, and the progress they detect can usually be explained by successful experimentation. That the "riddle of the universe" will always frustrate science does not discourage the expectation that each new effort may result in a significant discovery: all that is needed is the right combination of intellect, design, and tools.

This image of relentless movement toward wisdom has also been held by most geographers, even though the new of their enterprise has not always been more virtuous or more powerful than the old. Nor are the changing currents of geographical thought always manifestations of progress, for in addition to the refinements that are the hallmarks of a cumulative literature, the writings of geographers also reveal sporadic excitements that can be explained only by the fickleness of human nature. We embrace new ideas, use them for a while, then cast them aside. And the ideas thus abandoned may never have been disproved by objective standards; they may simply have been discarded because they were no longer fashionable. If one looks back, the landscape of American geography may seem littered with the detritus of abandoned tasks, but this impression is misleading, since our landscape is more accurately described as a scene composed of both relics and novelties.

The discussion that follows provides an account of the fate of a particular geographical concept that never achieved paradigmatic quality, even though it vied briefly for such a status by seemingly providing a historical determinism as a replacement for an increasingly unfashionable environmental determinism. Since there are few precedents for this kind of inquiry into the rise and fall of particular ideas, my effort, like John K. Wright's bibliobiography (1962), is experimental. I have selected sequent occupance for the reason that this concept attracted considerable attention in its time and because the period of its use can be delimited with reasonable precision. The term sequent occupance was introduced by Derwent Whittlesey in 1929; and although it is still used occasionally by American geographers (White, 1965; Krausse, 1971; Dillman, 1971), it had virtually disappeared from our literature by the end of the 1950s. During this period of approximately thirty years, sequent occupance evolved from novelty to relic without ever being subjected to critical review.

I. ORIGIN OF THE TERM

"Human occupance of area, like other biotic phenomena, carries within itself the seed of its own transformation." Whittlesey (1929, p. 162) offered both a model and a case study to support this declaration. The case study dealt with an area in New England where four stages of occupance, each expressive of internal evolution, could be identified. The first was an Indian stage, characterized by hunting and collecting in a forest. Next came a farming stage, during which gentle slopes were plowed and steep slopes were grazed. In the third stage the area was covered with second-growth forest, and occupance was limited to occasional grazing. On the basis of this progression, a fourth stage, when the forest would be cut periodically by nonresident owners, was hypothesized.

Whittlesey described each of these stages, or "generations of human occupance," as being "linked to its forebear and to its offspring," and their progression impressed him as being a consequence of internal forces analogous to the multiplication and destruction of living cells. Stage one was upset by technological innovations. As destruction of the forest commenced, stage two, farming, began. Farming, in turn, succumbed to decreasing yields, initiating a contemporary era of secondary forest.

But the model seemingly validated by this study was imperfect. As its author admitted, "the present era of idle land and renascent forest" was seen to represent "not a distinct mode of human occupance but a transitional period in which vestiges of the farming epoch linger on in the casual grazing of the margins, and in which an earnest of the epoch to come is offered by constantly meliorating tree growth." Moreover, although alterations evolving from the inherent character of a particular mode of occupance "follow a normal pattern and at length usher in a new and consequent mode," external forces are "likely to interfere with the normal course, altering either its direction or rate, or both." Hence normal sequences might be rare, or perhaps only ideal.

Whittlesey's caution about normal sequences might have checked the adoption of his model if geographers had not been conditioned by exposure to the Davisian conception of an ideal erosion cycle to believe that external or disruptive forces could be incorporated into schemes of internal evolution; and Whittlesey may have endorsed this notion in a later study of Ellsworth, Maine, when he referred to chorological rejuvenation produced by tourism (1931). In any case, the qualifications Whittlesey suggested at the end of his 1929 essay proved to be less important than his notion of an ideal sequence and the promise of prediction it offered.

The erosion cycle had already encouraged a generation of American geographers to think in such terms, and it seemed reasonable to suppose that these ideas could be transferred from the brown to the black lines on topographic maps (for an earlier attempt to make this transfer, see Rich, 1920). Moreover, historical investigations designed to produce generalizations of an evolutionist character had already been conducted by several geographers, and these studies seemed to offer an alternative to environmentalism (James, 1926, 1927, 1929, 1957). Yet, ironically, sequent occupance could also be welcomed as a way of avoiding the more emphatic historicism that was evident in the conception of geography as a discipline devoted to narrative study of the processes that transform natural landscapes into cultural landscapes. By presenting stages or cross sections rather than a more complete account of the development of landscapes, geographers could subordinate the chronological to the chorological and thus avoid the commitment to detailed historical and even prehistorical investigation that Sauer had recommended in 1925.

The adopters of sequent occupance may also have been encouraged by a desire to decorate their discipline with neologisms. Occupance was already well established in the American geographical literature, and the response to Whittlesey's phrase seems to have been generally favorable, although geographers, sensitive to the authority of their dictionaries, occasionally rendered it as "sequence of occupancy." In any case, contributors to the *Annals* offered these endorsements:

> It should soothe the ear, slip easily into the mind, and if possible define itself or almost define itself. Such a word is Mr. Whittlesey's "occupance." (J. Russell Smith, 1935, p. 18)

> We should ask before a new word is offered, "will this word by its fitness and its sound win friends?" Some terms are pleasant to the ear; I like Whittlesey's phrase "sequent occupance" and Trewartha's "confluence site." They have a pleasant sound and the advantage of meaning what their derivations suggest. These are two qualities that a new word ought to have; it should have an agreeable sound and it should define itself. (Whitbeck, 1934, p. 87)

II. APPLICATIONS OF THE IDEA

Was sequent occupance a new concept or merely a new term? Most of the ideas in Whittlesey's essay were already familiar to American geographers. Preston James recalls that the procedures and objectives later identified with sequent occupance were discussed during the annual spring field conferences held in the Midwest in the 1920s (personal communication, 4 June 1973; see also James, 1972). Nevertheless, several features of Whittlesey's presentation suggest more than a new label for a familiar enterprise. The assumptions that landscape evolution could be described in terms of diagnostic stages, that the progression of these stages could be predicted, and that variations from predicted or normal sequences would result from exogenous forces added up to a new generalization about landscape evolution and not merely to a reiteration of what had already been verified. Moreover, since these ideas were evident in physiography and plant ecology, sequent occupance is probably best described as an analogue model—that is, an attempt to transfer concepts from natural science to human geography. In short, sequent occupance was a heuristic device, and as such could endure only if it inspired work comparable in rationale to Whittlesey's New England study.

Such work seems to have been initiated almost immediately, for within two years Dodge presented a study of sequent occupance on an Illinois prairie (1931), and Thomas published a paper on sequent occupance in a section of St. Louis (1931). These studies seemed to confirm the idea that the interplay of landscape-building forces could be examined effectively by means of a series of temporal cross sections. Dodge described three stages: pioneer settlement along a tributary of the Illinois River, first expansion into the prairie, and completion of the settlement of the prairie. Thomas found five stages: pioneer, farmer, village, mining, and manufacturing.

Moreover, Thomas drew profiles representing typical landscape features during each of these stages and thus initiated a method of illustration that soon became a characteristic feature of sequent occupance studies. The use of this illustrative technique also implied belief that sequent occupance studies might be likened to the use of single frames from a motion picture to capture the essence of landscape evolution. This additional idea—that the stages, or cross sections, could be regarded as portraits—was endorsed by Whittlesey's latter assertion that "stages of occupance" should refer to "an epoch during which the human occupance of an area remains constant in its fundamental aspects" (1933, p. 651n.).

The several ideas implicit in the introduction and early application of sequent occupance soon enjoyed currency among geographers. During the Conference on Regions, which highlighted the 1935 meeting of the Association of American Geographers, Robert Hall declared that "more and more emphasis must be directed toward cultural succession"; and the Davisian formula—structure–process–stage—was given human-geographical connotation by George Cressey's assertion that cultural landscape "equals the fundament, plus the culture, plus the sequent occupance or succession" (1935, pp. 219 and 133). In a later conference on cultural geography, Hall recommended that an attempt be made to "layer occupance upon occupance in much the same manner as the geologist studies stratifications" (1927, p. 168).

The enthusiasm evident in these remarks was more emphatically expressed in 1934, when James used sequent occupance as one of the organizing themes in a textbook. His major theme, derived from Siegfried Passarge's concept of *Landschaftsgürtel,* was the division of the earth into bioclimatic realms. The several strategies followed by James in his account of the human geography of these realms included sequent occupance or "the concept of a succession of cultures more or less separate in time, and each marking the face of the earth in accordance with its own mode" (1934, p. 33). Thus he described three stages of occupance for the Mediterranean scrub-forest lands: first, a pioneer stage, characterized by "scattered or patchy distribution of settlements in the midst of undeveloped lands"; then a stage of elaboration "when the outline of settlement is filled in and begins to show a closer and closer relationship to the underlying qualities of the land"; finally, a climax stage, when "adjustment becomes nicer and more detailed" and "each significant change in the character of the soil, the surface, and drainage, or any other feature of the land is reflected by a change in the way that land is used" (1934, pp. 126–27).

The most striking evidence of the early popularity of sequent occupance was its use in doctoral dissertations. The first such use seems to have been at the University of Michigan, where studies by Glendinning (1935), Kendall (1935), and Davis (1936) contain brief sections on the sequential development of settlement and land use. Later Meyer (1936) presented an account of the "sequent occupance forms and functions" of the Kankakee Marsh of northern Indiana and Illinois, and Alfred Wright (1936) offered a study of three periods in the evolution of the industrial geography of the middle Miami Valley of Ohio. Meyer's work was especially important in that it offered a detailed description of sequent occupance and because the silhouettes illustrating the several stages of his study brought the idea of snapshot historical geography to effective realization.

Still, the fact that sequent occupance probably attained a climax in Meyer's work of 1936 did not discourage further application of the ideas and research strategies embodied in the concept. For example, sixteen out of sixty-four dissertations completed at the University of Chicago between 1936 and 1956 reveal explicit or implicit adoption. Six of these studies present the sequential development of settlement and land use as their major theme (Glasgow, 1939; Carlson, 1940; Richards, 1948; Matthews, 1949; Schrode, 1948; McGaugh, 1950), and the other ten offer cross sections for particular dates or descriptions of longer intervals as a contribution to understanding contemporary scenes (Espenshade, 1944; Nelson, 1949; Wheeler, 1950; Sorensen, 1951; Thomsen, 1953; Erickson, 1953; Christensen, 1956). The group of six apply sequent occupance as an organizing theme, whereas the other ten adopt the ideal of the cross section as a convenient way to present necessary historical background.

Historical geography had of course been practiced by

University of Chicago students since the time of Sauer (1915) and Parkins (1918) and was presented, largely in terms of an evolutionary view of human adjustment to varied environments, in the teaching of Barrows (1933; see Koelsch, 1962). Moreover, two dissertations completed at Chicago employ terminology (youth, adolescence, maturity, old age) that belongs to a parallel adventure in geographical ideology (Minton, 1939; Schockel, 1947; see also Taylor, 1945; and Van Valkenburg, 1944). Therefore, many of the Chicago students who adopted frameworks or approaches that can be described as sequent occupance may only have glanced at Whittlesey's essay of 1929 or may have received its message indirectly. The distinction here is not unlike that between the Turner thesis and historical interpretations that are merely Turnerian (Gulley, 1959). In any case, the use of sequent occupance at the University of Chicago developed slowly but persistently after 1939, reached a peak in 1948, and then declined until the final study of this type was presented in 1956. This evidence of rise and decline offers ample opportunity for speculation, but first we need to know more about the rationale and accomplishment of the many geographers who tried to recreate a succession of geographies. Sequent occupance studies may be grouped according to their organization and presentation. Economic and land-use criteria were used most commonly in the determination of stages—that is, subsistence farming stage, lumbering stage, mining stage, industrial stage, and so on—but few studies displayed consistency in this regard. For example, Ackerman (1941) defined five periods in the sequent occupance of a suburban Boston community: aboriginal Indian, colonial subsistence, dairying and manufacturing, rural depopulation, and intensive truck farming and residential property expansion. Similarly, in his study of the High Plains of Michigan, Davis (1936) referred to an Indian period, a period of French fur trader and Jesuit missionary activity, a period of British control, and a period of American forest clearing and pioneer agriculture. As the use of ethnic, economic, demographic, and political labels in these studies implies, different stages were defined by different criteria.

Occupance periods, or stages, were usually given ethnic or political labels (for example, Indian, British, American) but were described in terms of land use, settlement, or communication (pioneer farming, urbanization, impact of the railroad). In other words, definitive expressions were coined to fit the character of each stage, and little effort was made to coordinate such expressions with those used to describe previous stages. Indeed, the literature of sequent occupance includes many examples of cross sections that differ so notably in emphasis that the connection among sections is difficult to establish. Since each stage usually introduced a new directing force, the design of sequent occupance studies often did not allow for, or proceed from, a persisting analysis of trends.

On the other hand, at least a few contributions to the

sequent occupance literature placed stress on a particular process or landscape feature. Nicholson's treatment of the settlement of southwestern Ontario (1952), where "development could be traced in several stages following changes in the mode of transportation," evokes processes, whereas Dodge's (1931) assertion of "the fundamental importance of drainage" in the sequent occupance of an Illinois prairie reflects the primacy of landscape features. A third class of sequent occupance studies, devoted to more specific aspects of occupation, evinced consistency in the definition of stages. Thus, McCune's study of the dominance of different plantation crops in Ceylon produced a statistically defined sequence consisting of a cinnamon period, a coffee period, a tea period, and so on (1949). Faced with a more complex transition in the agricultural history of southern California, Gregor suggested a subsistence stage, a grain stage, a mixed-crop stage, and a specialized fruit and nut stage (1953).

Effort might expectably have been devoted to the selection of the times to be studied. Yet many studies, especially those employing sequent occupance as historical background, offered information for purely arbitrary intervals. Thus, Pounds presented descriptions of the Ruhr at 1800, 1850, and 1900 "as a prelude to examination of the Ruhr in the mid-twentieth century" (1952, p. 15). Similarly, Sorensen offered descriptive cross sections of Springfield, Illinois, in 1855 and 1890, as "a background for at least a casual investigation of the genesis of selected patterns in the contemporary city" (1951, p. 45).

III. REEXAMINATIONS

Students of sequent occupance allowed themselves a wide range of options. They could take account of several elements of material or nonmaterial culture or deal with only a particular variable, such as population density, transportation facilities, or land use. A further option concerned the intervals of time within which descriptions were supposed to be valid—a year, a decade, a century, or even longer intervals, such as, for example, the Indian, Spanish, and American periods in the history of California. Finally, Whittlesey's concept was extended from a small part of New England to a number of areas in the United States and even to one of the world's major natural realms.

The fact that most of these options were evident soon after the introduction of sequent occupance should have encouraged both a serious reexamination of Whittlesey's initial assumptions and a critical review of the derivative literature. It is surprising that this did not happen, since geographers enjoyed excellent communication in the 1930s. A possible key to this neglect is found in Stanley Dodge's suggestion that sequent occupance might be merely "a useful expedient"; yet he also suggested a three-part definition that would have imposed a measure of discipline on the rapidly expanding literature.

> If "sequent occupance" does not become merely a phrase to cover a vague idea, three things must be involved in its meaning, (1) the idea of sequence, of time, (2) the idea of the thing or things occupying, and (3) that of the thing or things occupied. (Dodge, 1931, p. 205)

This suggestion, published only two years after the appearance of Whittlesey's essay, does not appear to have influenced later studies. Nor was much heed paid to the complaint of Richard Dodge, that he could find "no agreement as to ways in which aspects of former occupance, reflected in the cultural features of the landscape, should be classified" (1938, p. 233).

Whittlesey himself made no substantial contribution to the evolving lore of sequent occupance, although his comments after 1929 indicate a retreat from the confidence displayed in his initial essay. Indeed, as already indicated, by 1933 he was ready to assert that the stage of occupance should be defined in reference to a relatively quiet interval: "an epoch during which the human occupance of an area remains constant in its fundamental aspects" (p. 451). The same essentially ahistorical tone appears also in Whittlesey's contribution to the conference on cultural geography during the 1936 meeting of the Association of American Geographers, when he held that "sequent occupance implies that what has existed in the past is our concern only if it has left vestiges and so exists also, in effect, in the present" (1937, pp. 168–70).

This was a far cry from Whittlesey's first contention that studies of sequent occupance would lead to both retrospective and predictive generalizations. Yet even in 1929 he was cautious about "normal sequences" and included disclaimers in his endorsement of biological and physiographical analogies. It is not surprising, therefore, that Whittlesey's most emphatic reassessment of sequent occupance was based on methodological grounds. Indeed, only in this prescriptive context can the question posed in his presidential address of 1944 be understood: "Is there a solution for the puzzle of writing incontestable geography that also incorporates the chain of events necessary to understand fully the geography of the present day?" (1945, p. 32). Viewed in this light, sequent occupance seemed to offer geographers an opportunity to include both historical and geographical dimensions and yet still practice their craft according to Hartshorne's 1939 guidelines for the separation of geography and history. The idea of employing static cross sections to avoid confusions of historical and geographical objectives—an idea Hartshorne had adopted from Hettner (1905)—was congenial to the introspective spirit of American geographers in the 1940s. Such studies were, in Whittlesey's words, "indubitably geographic and could be judged with propriety by geographers" (1945, p. 38). Yet the assumptions that the primary and logical emphasis of geographical research should be on the present day encouraged a skeptical attitude toward even this restricted or subordinate view of historical geography. As Whittlesey cautioned,

> No series presenting the geography of consecutive earlier periods has yet been prepared for any large area. When done, the series would presumably recreate the geographic landscapes for all the epochs for which records can be found. From studies which have been made, it appears that much of the latter would have no relevance to present-day geography of the area. The sequence of wanted items might be culled from the facts presented, but it would not leap to the reader's eye and mind. (1945, pp. 31–32)

IV. DECLINE

The fact that Whittlesey felt obliged to question the relevance of his concept after a decade and a half of its extensive employment suggests that the future of sequent occupance was in doubt among geographers concerned primarily with the present day. Moreover, the restricted conception of the scope and objective of sequent occupance studies that was implicit in the methodology of Hettner and Hartshorne precluded any substantial use of the concept by geographers who sought support and inspiration from the neighboring fields of history and anthropology. Too grand in its initial form to attract such scholars, sequent occupance became too restricted for their taste after it had been redefined to conform to an ahistorical methodology.

In addition, the assumption that sequent occupance might be a route to a new determinism in geography clearly was not taken seriously. As James indicated in his textbook,

> There is, of course, nothing inevitable about such a sequence of stages. In some cases areas may stagnate almost indefinitely in the stage of elaboration. In others a region may advance rapidly from a pioneer stage to the development of a number of areas of climax. . . . In still others, which may have reached a climax, events may cause a retrogression. (James, 1934, p. 127)

Studies employing sequent occupance as a convenient device for the presentation of historical background displayed even more skepticism. Thus Pounds claimed "no particular importance" for the dates or periods of his historical surveys, and he did not "altogether refrain from tracing one pattern from that which preceded it" (1952, p. 15). Wheeler's disclaimer is even more emphatic:

> Although these dates are convenient points of reference, it must be emphasized that they do not represent sharp breaks in the sequence of development. Each period has tended to merge gradually with the next, and many of the institutions, attitudes, and cultural features of particular periods have survived in succeeding periods, or even to the present time. (1950, p. 68)

More generally, it is probably safe to assume that sequent occupance entered American geographical lore during a phase of antithesis rather than thesis, when the prospect of historical determinism replacing environmental determinism had little appeal. It began to fade away at the time of a new impetus toward functional and generic theses. When regional geographers began to embrace the several concepts that would turn attention toward economic and urban geography, even the halfhearted historical interest that they took in sequent occupance could be dismissed as antiquarianism.

As for the few geographers who developed an explicit commitment to historical geography, the prospect of writing several past geographies, especially for arbitrarily or vaguely defined periods, held little appeal. To be sure, comprehensive reconstructions had been achieved for particular past periods, notably by Brown (1943), but the sequent occupance approach required nothing less than a succession of such reconstructions. Viewed in Clark's terms as the study of "changing geographies" or the "geography of change" (1960, p. 613), the snapshot approach of sequent occupance had little utility for historical geography. Even if conceived as merely the geography of a past period of time, historical geography framed by reference to "an epoch when the human occupance of an area remains constant in its fundamental aspects" would permit only implicit consideration of the forces that produce change. Not surprisingly, therefore, sequent occupance receives relatively little attention in surveys of historical geography (Clark, 1954; Merrens, 1965; Prince, 1971).

Sequent occupance presented corresponding difficulties to cultural geographies, especially to those associated with Sauer at the University of California, Berkeley. The emphasis on independent development in situ could not be adopted by scholars who had placed major emphasis on migration and diffusion. Several key sequent occupance notions (supercession, stage, uniformity) were in conflict with other notions (transition, overlap, pluralism) that figured prominently in cultural geography. Cultural geographers with an anthropological perspective could dismiss sequent occupance as a form of the discredited notion of "unilinear evolution" (Steward, 1955). And geographers concerned primarily with material culture could not be excited by an approach that offered effects without causes and demanded acceptance of the premise that process is implicit in stage. Indeed, for all its defects, the Turner thesis was probably a more useful approach, for it entailed recognition of the importance of movement. Turner, in his address of 1893, invited his listeners to "stand at the Cumberland Gap and watch the procession of civilization, marching single file—the buffalo following the trail to the salt springs, the Indian, the fur-trader and hunter, the cattle raiser, the pioneer farmer—and the frontier has passed by" (Turner, 1894). In contrast, Whittlesey might be said to have invited his readers to stand on the edge of a small area in New England and watch it transform itself.

Cultural geographers felt also that the sequent occupance focus on relatively static intervals, with its concomitant use of profiles or silhouettes, tended to mask or misrepresent actual sequences. Kniffen's 1951 study of transportation development in the lower Mississippi Valley exemplified this criticism effectively, for each of the modes of transport that he described, beginning with pirogues and canoes and ending with airplanes, extended well into the period of its successors. Horses on trails and roads overlapped the use of flats and keelboats, river steamboats, and towboats, and extended even into the era of automobiles on highways. Based on intensity of use, the horseback stage was in large measure also a riverboat stage; the transition from rail to highway dominance was likewise gradual.

Similarly, the effectiveness of Broek's famous 1932 study of the Santa Clara Valley is largely attributable to his skill in combining four descriptive cross sections (the primitive landscape, the landscape of Spanish-Mexican times, the landscape of the early American period, and the present landscape) with three accounts of the social and economic forces that fostered landscape change (see also Darby, 1960, 1973). Broek's study, in dealing with overlap as well as succession, and with turbulent as well as relatively static intervals, presented a more complete picture of landscape evolution than a series of static cross sections alone could have offered.

Since cross sections had doubtful merit in historical research, the methodological value of sequent occupance rested on its one testable hypothesis: the dubious assumption that change results largely from internal forces. But sequent occupance studies in fact usually demonstrated the importance of external forces and diffusion. Such studies might have led to descriptive or classificatory generalizations, but even these possibilities were frustrated by shifts of emphasis. Hence it is not an exaggeration to suggest that the fate of sequent occupance was decided when those who adopted it failed to take account of the suggestions made by Stanley Dodge in 1931 and Richard Dodge in 1934. Had their suggestions been accepted, or at least kept in mind, it might have been possible to compare similar stages of occupance in different regions or the effects of different processes on landscape evolution. In fact, by accepting any one of a wide range of options in the design of their studies, the adopters of sequent occupance produced a literature that was additive rather than cumulative. James' assertion in 1948 that "explanatory studies in sequent occupance add up to something" (p. 274) in effect begged two important questions: whether such studies had indeed been explanatory, and whether the something they added up to could provide a basis for a comparative study.

As a new style of regional description, sequent occupance did inspire some works appropriate to the humanistic and artistic traditions of geography—for example, Higbee, 1952—and the rare studies focused on a particular process or landscape feature were effective by the standards applied to regional geography (Trewartha, 1940; Hall, 1937). But the formidable historiographic problems that sequent occupance entailed could not be solved by the scholarship most commonly employed. Indeed, as suggested above, sequent occupance reached an early culmination in Meyer's work of 1936, since subsequent studies by American geographers (excepting those by Meyer himself) were not of comparable quality. From its initial status as a model, sequent occupance retrogressed to become merely an organizational device in regional studies. Even in Whittlesey's later work, sequent occupance was only a subordinate theme in a program that led to his proclamation of the "compage" as geography's most compelling challenge (1956).

These comments may provide an adequate explanation for the demise of sequent occupance, although the fact that the concept was not subjected to such criticism while it was in widespread use suggests that the fickleness of human nature, mentioned earlier as a necessary postulate in the history of geography, may have figured prominently in its decline. The enthusiasm shown toward Whittlesey's concept in the 1930s offers an additional illustration of human nature in geography, for it is surely significant that most people readily identify stages in their own lives, and that attempts to impose such schemes on the history of civilization can be traced into antiquity.

This final generalization is of course merely speculative: although we can hope to detect trends in the published lore of an academic discipline, it is difficult to be confident about subtle currents of influence and rationale. In any case, the issue just raised can be rephrased as a question: Was sequent occupance a predictable or inevitable adventure for American geographers? This question has to be answered affirmatively, if we endorse the premise that most scholars find it difficult to accept purposelessness or disorder. Neither history nor geography can be written indiscriminately; and since time cannot be comprehended as a continuum, the search for order in historical geography must entail sequential arrangement. Therefore, when geographers embraced the idea of historical explanation as an alternative to environmental explanation, sequent occupance had instant appeal. Nor is geography unique in this respect; for concern about causes and effects is a necessary part of any scholarly enterprise, and extension of this concern requires the use of temporal stepping-stones. It is not surprising, therefore, that each of the groups devoted to the study of mankind, even economists—for example, Rostow, 1960—has at least flirted with stages, cycles, and inevitable progressions; nor is it surprising that attempts to offer such schemes as an explanation of human destiny—from Hesiod and Varro to Marx and Spengler and on to the "universal historians" of our time—figure prominently in intellectual history (for an elaboration on this theme, see Lovejoy and Boas, 1953, and Kramer, 1967). Whittlesey may be placed in this company, for, although the scale of his generalizations was less than universal, he seems clearly to have believed, if only for a moment, that he had found a key to cultural evolution.

V. SUMMARY

It is often claimed that the character of current geographical thought exhibits the contrast between the old and the new. Yet no decade in the history of geography has been devoid of this contrast. Moreover, the assertion that the contrast of old and new is or has been a prominent feature of our lore begs a perplexing question, for one is always tempted to ask whether the changing fashions of geography reflect the refinement of a science or simply the fickleness of human nature. The fact that we are often unable to answer this question means that a good case can be made for experimentation in the history of geography.

In 1962 John K. Wright offered one such experiment by treating the career of Ellen Semple's "Influences of Geographic Environment" as a case study in the history of American geography. I have attempted here to examine the career of sequent occupance, which was introduced to American geographers in the 1920s, was widely used in both teaching and research in the 1930s and 1940s, and then faded from our scene in the 1950s. Sequent occupance probably became popular because it was an analogue model, built in emulation of the erosion cycle and the concepts of succession and climax in plant ecology. Sequent occupance seemed also to promise a historical determinism to replace environmental determinism. But it was Whittlesey's case study, rather than his highly tentative generalizations about landscape evolution, that encouraged the proliferation of sequent occupance studies, and the large literature thus inspired proved to be additive rather than cumulative.

Since sequent occupance was not subjected to serious criticism while it was in widespread use, we can only speculate on the reasons for its eventual decline. Speculations that seem plausible include the influence of ahistorical methodology in the decade after 1939 and the formidable difficulties inherent in the recreation of a succession of geographies. In addition, few geographers interested in the past viewed with favor a conceptual framework that emphasized local development rather than diffusion and that required acceptance of the Davisian assumption that process is implicit in stage. But since the weaknesses and inherent difficulties of sequent occupance were seldom mentioned, its abandonment in the 1950s may have been prompted by a general but unstated realization that the program, which seemed exciting in the 1930s, had become a routine and redundant exercise a generation later. If so, the life cycle of sequent occupance—from initiation through adoption and application to neglect—may best be explained, in Wright's terms, as a manifestation of human nature in geography.

2 GEOGRAPHICAL THEORIES OF SOCIAL CHANGE

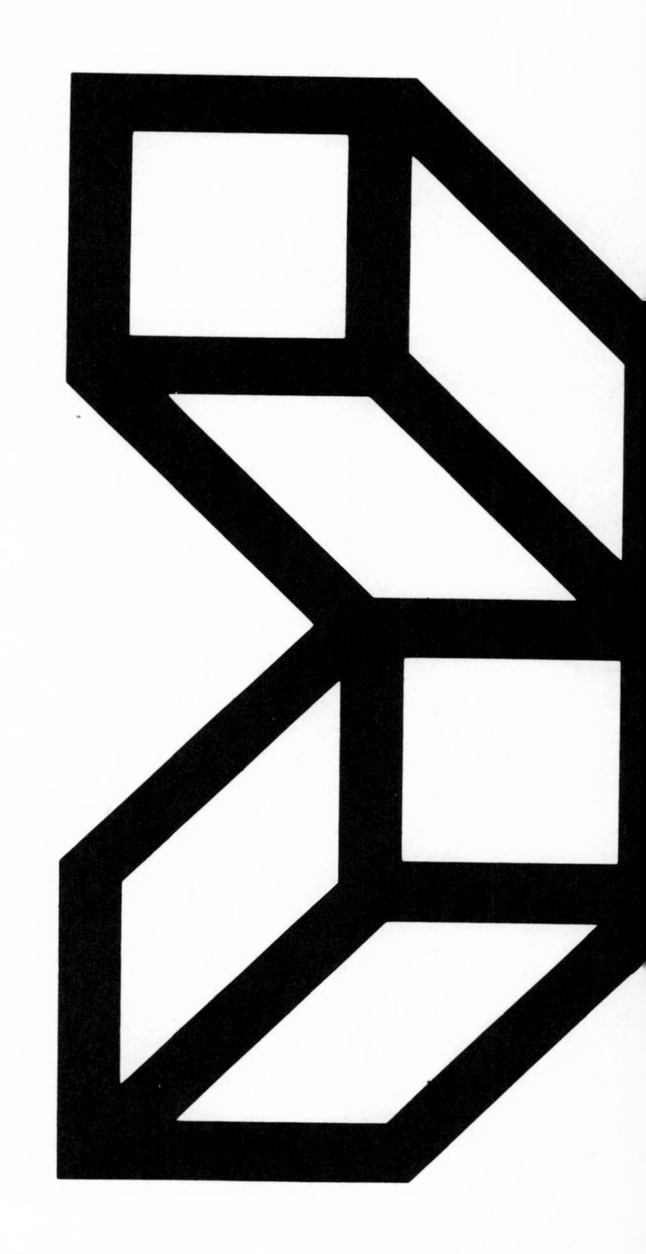

Brian J. L. Berry
Harvard University

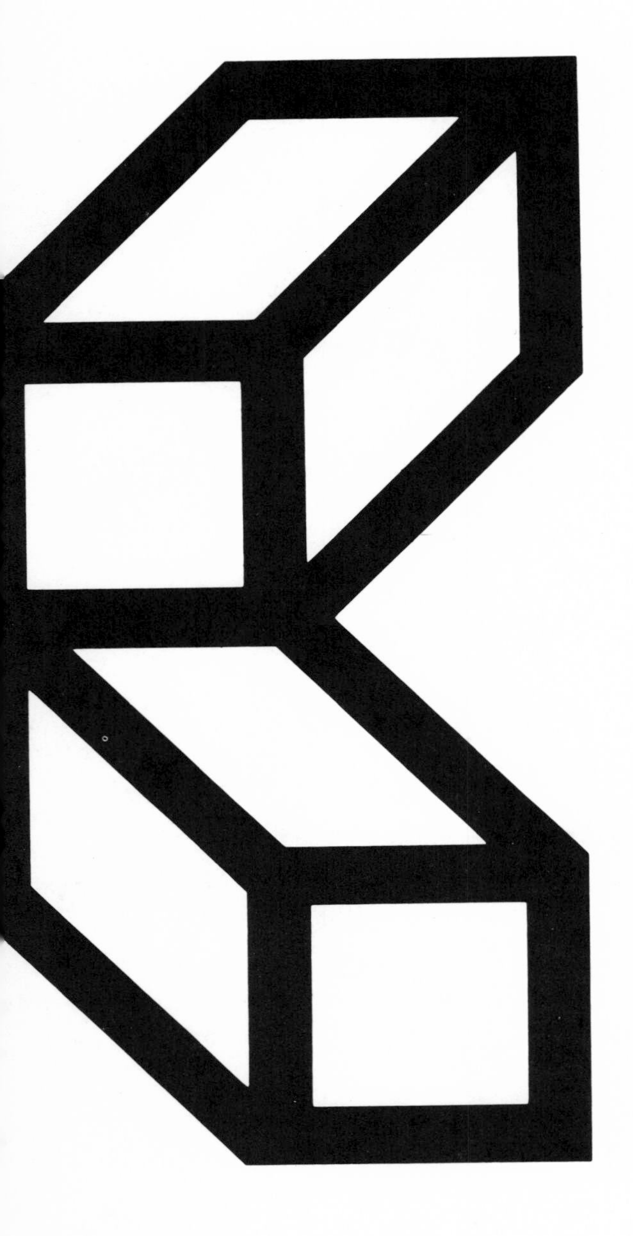

GEOGRAPHICAL THEORIES OF SOCIAL CHANGE

If the rise and fall of sequent occupance is an example of human nature working in geography to propose and later to discard ideas as unfashionable, then shifting geographical theories of social change exemplify the larger paradigm changes that have characterized the scientific progress of the field. In such shifts there is both continuity and change, as Kuhn recognized. Paradigm changes have caused geographers to see the world in new and different ways—to identify facts differently, to construct different patterns of explanation, and to pose new types of hypotheses for testing, evaluation, and integration into more elaborate theoretical structures. Yet in one paradigm after another certain themes have reappeared.

I. BACKGROUND IN INTELLECTUAL HISTORY

This is illustrated no more graphically than by geographical theories of social change, which have intellectual roots extending from antiquity in men's speculation about the habitable earth and their relations to it, and a diversity arising from the mosaic quality of modern geography which results from the many points of origin of the field. Yet they have also moved from one paradigm to another, and in the last decade they have been extremely dynamic, for at the very time that geographic thinking has moved beyond the demythologizing of nature through the acknowledgment of man's agency in change to reach the frontiers of cybernetic control and intentionality, it has been confronted by a resurrection of the questions that concerned the ancients by the created awareness of a new environmental crisis.

The Classical Roots

The traditionally held view is that geography is concerned with giving man an orderly description of his world and his place in it, a task complicated by the bewildering phenomenological diversity involved. Nevertheless, in the history of Western thought men have persistently asked three questions concerning the habitable earth and their relationship to it (Glacken, 1967):

> Is the earth, which is obviously a fit enviroment for man and other organic life, a purposefully made creation?
>
> Have its climates, its relief, the configuration of its continents influenced the moral and social nature of individuals, and have they had an influence in moulding the character and nature of human culture?

In his long tenure on earth, in what manner has man changed it from its hypothetical pristine condition?

From the time of the Greeks, answers to these questions have been advanced so frequently that Glacken proposes that they be restated as three general ideas: the idea of a designed earth; the idea of environmental influence; and the idea of man as a geographic agent. Each idea has implicitly contained a particular geographic concept of change.

The idea of a designed earth owed much to mythology, theology, and philosophy, and led to intersections of geographic and theological thinking at crucial points of human curiosity. If we seek after the nature of God, it was thought, we must consider the nature of man and the earth; and if we look at the earth, questions of divine purpose in its creation and of the role of mankind in it arise. In Stoic pantheism the idea of a deity and of nature were one; in Christian theology they supplemented and reinforced each other. From the beginnings, the idea of unity and harmony in the universe frequently became part of the idea of purposefulness in the creation, an idea that finally was articulated in the teleology of Plato and Aristotle, enriched by the Stoics, imprinted on the Christian Middle Ages in Judeo-Christian theology as the idea of the earth as a planned abode for man, and carried through into Renaissance thought.

The idea of environmental influence was based on pharmaceutical lore, medicine, and weather observation, relying on comparisons of environmental factors with the different individuals and peoples characteristic of these environments. In the Hellenistic period it appeared in the dogged Epicurean resistance to Stoic teleology. While sharing the notion of unity and harmony in nature, the Epicureans substituted nature as a creatrix for divine planning. Ideas of environmental influence and adaptation reappeared throughout the Middle Ages as explanations of racial and cultural differences, whether in the Western writings of St. Thomas Aquinas or the work of Muslim scholars like Ibn Khaldūn (Kimble, 1938; Wright, 1925), and they grew to an ascendancy in the eighteenth and nineteenth centuries.

The idea of man as a geographic agent emerged from considering the plans, skills, and activities of everyday life. In Greek thought there was an awareness of man as a creator of order. In Christian ideologies man could still function as a finisher of the creation, bringing order into nature and unlocking its divinely endowed properties; man, created in God's image, had by God's grace dominion over all nature. These ideas became sharper during the Renaissance, leading to the expression by Bacon and Descartes of purposive human control of nature and to increasing emphasis on human society and its accomplishments and to the possibility of improving society by the purposive application of scientific law to the needs of food, housing, and the like. But already, presaging later interests in conservation and the new prophets of environmental failure, an antiphonal idea appears—that men often make changes in nature that are reckless as far as long-term trends are concerned and purposive only for narrower self-serving ends.

The Eighteenth Century

All three ideas were discussed with thoroughness and penetration during the eighteenth century, as man sought new knowledge about societal bonds, tradition, national character, the environmental influences affecting the lives of individuals and of nations, and the complexities of human history. They were all therefore present as the foundations were laid for the emergence of the social sciences during the nineteenth century.

Linnaeus and Buffon provided new insights into nature. Travelers speculated about stages of human development and the influences determining the character of peoples. Hume and Kant criticized teleological judgment. Dramatically, in *De l'Esprit des Loix* Montesquieu revived environmental theories, summarizing increasing speculation about the relative merits of the ancients and the moderns, differences between the races, concepts of disease and public health, and explanation of national character, only to be criticized by Voltaire because he (Voltaire) felt that there was no scope for bad laws and government to be changed if they were held to be products of the environment. Malthus propounded an environmental theory of population growth. Humboldt and Buffon added observations about climatic influences on population, although Buffon reemphasized the notion of man as an agent of geographic change.

This maelstrom of ideas, combined with nineteenth century industrialization of the Western world, new theories of the origin and evolution of life and human culture, and growing specialization of knowledge, was the framework within which modern geography emerged and the discipline articulated its first explicit theories of social change.

The Emergence of Modern Geography

School geography had, during the eighteenth century, in contrast to the general concerns of natural philosophers about man and nature, become narrowly focused on the summary of facts about countries and cities. The most widely used tests were those of Bernhardus Varenius, written in the mid-seventeenth century in response to requests for practical information by Amsterdam merchants, and similar volumes written for British and American audiences: Patrick Gordon's *Geography Anatomized,* which had twenty editions in England from 1693 to 1728; William Guthrie's *New System of Modern Geography* (1770, revised through 1843); and Jedidiah Morse's *Geography Made Easy* and subsequent texts, the world's first financially rewarding educational publications (James, 1969).

Nevertheless, new geographies were emerging. In 1762,

as part of his revolt against traditional methods of educating children, Jean Jacques Rousseau had outlined a new method of teaching that would enable children to develop their individual capacities. At the same time, the Swiss educator Johann Pestalozzi argued that clear thinking was dependent on accurate observation—that words could have no meaning unless directly related to something directly perceived by the senses. In 1784, a German schoolmaster named Christian G. Salzmann, excited by these ideas, took a five-year old boy unexposed to traditional methods of education, to experiment with the new methods (James, 1969).

This boy was Carl Ritter, and he developed a new scientific approach to geography in 1817. He emphasized the unity of man and nature. But every detailed observation, said Ritter, must be related to the general laws that govern the operation of processes on the earth, and he chose, therefore, to emphasize the physical earth. The ultimate explanation of the general laws he held to be one of divine purpose, and the irrevocable working of processes was towards divinely endowed, but currently latent, ideal or ultimate states.

Ritter's teleology and physical geography were carried to the New World in 1848 by Arnold Guyot, who became professor of geography and geology at Princeton University. Guyot and Matthew Fontaine Maury, whose school text on physical geography had been adopted by 60 percent of the states by 1875, did much to foster the emergence of geography as a university-level discipline out of geology—a consequence of the interest among physiographers in the influence of natural environment on mankind. The first university department of geography, established at the University of Chicago in 1903, stated as its purpose "to occupy the ground intermediate between geology and climatology on the one hand and history, sociology, political economy and biology on the other." Meanwhile, continuing to evolve out of the eighteenth-century cosmography that provided a more general setting for Ritter's work, geography in Germany emerged not as a specialized discipline but as a comprehensive earth science (*Erdkunde*). In France, geography grew out of history; and in Britain the first university scholars who identified themselves as geographers were concerned with the contributions they might make to clarification of Britain's role in the world and the management of the empire.

With multiple ideas and multiple origins, modern geography could rightly be characterized as a "mosaic within a mosaic" (Mikesell, 1969), a subject with manifold interdisciplinary connections, consuming ideas from many sources. A first round of university expansion took place at the beginning of the twentieth century, and at this time its major professional associations were established and its journals were initiated. At this time, too, the first paradigm of modern geography crystallized. Since then, geography has moved away from its original home in natural science or the humanities, however, seeking first to function as a bridge between nature and culture and then as an autonomous but pluralistic enterprise concerned with man-land relations area studies, man's spatial organization, and physical geography as a component of earth science.

II. THE FIRST PARADIGM OF MODERN GEOGRAPHY: ENVIRONMENTAL DETERMINISM (SOROKIN'S "GEOGRAPHICAL THEORY")

The point of departure for this subsequent pluralism was a particular paradigm that provided an initial focus for the field. It was so strong that when, in 1928, Pitirim A. Sorokin published his masterful review and critique, *Contemporary Sociological Theories Through the First Quarter of the Twentieth Century,* it led him to devote one-sixth of his study to it, and to call it *the* geographical theory.

"Almost since the beginning of man's history," he wrote, "it has been known that the characteristics, behavior, social organization, social processes and historical destiny of a society depend upon the geographical environment" (Sorokin, 1928, p. 99). Yet after his review of the environmentalist hypotheses, he reflected that "At the bginning of a study of these theories one is impressed by their brilliancy and originality; continuing the study one is perplexed and bewildered by their contradiction and vagueness; and finally he is lost in the sea of these theories not knowing what in them is valid, and what is wrong or doubtful. This explains why the primary need in this field at the moment consists not so much in a formulation of a new geographical theory or a new correlation between geographical factors and social phenomena as in a most rigorous analysis and shifting [*sic*] of what is valid and what is childish in these numerous hypotheses" (Sorokin, p. 101). But this rigorous analysis never came. By the time Sorokin summarized it, geography had simply disapproved (rather than disproved) the environmentalist thesis and moved on to new topics (Wagner and Mikesell, 1962).

The basic environmentalist idea was expressed no more clearly than by the first president of the Association of American Geographers, William Morris Davis, in his 1906 presidential address: "Any statement is of geographic quality if it contains . . . some relation between an element of inorganic control and one of organic response" (Davis, 1906, p. 71). Davis accepted the Ritter-Guyot explanatory method because it provided more general notions than the earlier body of unconnected facts, but replaced the underlying deism by the causal notions of environmentalism. Human society, he argued, was an organism which survived by adjustment to the physical environment; the nature of its growth was thus environmentally prescribed. This form of Social Darwinism he derived from the writings of Herbert Spencer (Herbst, 1961).

Davis was a tireless scholar who sought for many years to impose uniformity of environmentalist concept on American geographic education, receiving stimulus from widespread acceptance of the Darwinian hypothesis and the increased codification of knowledge about the environmental, racial, and cultural differences existing in the world. But the extremes of the view of nature as creatrix which he, Ellen Churchill Semple, Albert Perry Brigham, Ellsworth Huntington, and Griffith Taylor propounded were already being ameliorated by those who accepted the idea of environmental *influences* but rejected the notion of ultimate causes. Dominating scholars such as Frederic Le Play, who developed the first really scientific method of the study and analysis of social phenomena (Sorokin, 1928), had already correlated place with type of work, forms of property, type of family and superfamily institutions and associations, and many social processes. He emphasized, however, the mutual interdependence of place, economy, and culture. This work fed back and forth with that of the French geographer Vidal de la Blache as he formulated his *Principles of Human Geography* (1926), which in turn related back to Friedrich Ratzel's *Anthropogeographie* (1891), the basis of much emerging geography in Germany.

Vidal saw the dominant idea in all geographical progress as that of terrestrial unity. "The conception of the earth as a whole, whose parts are coordinated, where phenomena follow a definite sequence and obey general laws to which particular cases are related" and, paraphrasing Ratzel, "the phenomena of human geography are . . . everywhere related to the environment, itself the creature of a combination of physical conditions" (pp. 6–7). Haeckel's concept of ecology was accepted, namely (p. 9), "the correlations between all organism living together in one and the same locality and their adaptation to their surroundings" and (p. 10) "Every region is a domain where many dissimilar things, artifically brought together, have subsequently adapted themselves to a common existence," although he disavowed the extremes of the environmentalist position (p. 12), whereby "the influences of environment are seen only through masses of historical events which enshroud them."

Ratzel's work in volume 1 of *Anthropogeographie* was brought to American audiences by Ellen Churchill Semple in *The Influences of Geographic Environment* (1911). She wrote: "Man is a product of the earth's surface. This means not merely that he is a child of the earth, dust of her dust; but that the earth has mothered him, fed him, set him tasks, directed his thoughts, confronted him with difficulties that have strengthened his body and sharpened his wits, given him his problems of navigation or irrigation, and at the same time whispered hints for their solution" (p. 1). It remained for the critics of environmentalism, such as Carl O. Sauer, to carry through the implications of Ratzel's work, in volume 2, on cultural diffusion, however.

Sorokin effectively summarized the ideas of the environmentalists. Among the phenomena they explained by environmental differences were: population numbers, distribution, and density; housing types; road location; clothing; the amount of wealth produced and owned by a society (T. H. Buckle's theory that the history of wealth in its early stages will depend almost entirely on soil and climate), leading to the idea that all brilliant and wealthy civilizations of early times have occurred in "favorable" natural environments, whereas "unfavorable" climates, inaccessible and isolated areas, breed backwardness and savagery; the location and nature of industry; business cycles and the rhythms of economic life (Ellsworth Huntington's[1919] meteorological theory of business cycles, as elaborated by H. L. Moore [1923]); race (Buckle's concept that differences among societies in bodies, minds, social organization, and historical destiny are environmentally determined, extended by the work of Ratzel and Semple, and the geographer-turned-anthropologist Franz Boas—although Boas extended his concept of environment to include the social environment too); health, energy, physical and mental efficiency (Huntington's investigations); suicide; insanity; crime; birth, death, and marriage rates, religion, art, and literature; social and political organization of society (for example, Ratzel's theories of the state, in which states with large territories possess a spirit of expansion and militarism, an optimism and youthfulness, and a psychology of growth, with less social and racial conflicts than in small territories, which are characterized by pessimism, earlier nationalism, stagnation, and an absence of virility).

The leading environmental theories of social change were those of *Equatorial Drift* and the *Northward Course of Civilization.* The theory of the northward course of civilization, articulated most clearly by Gilfillan (1920) was that "the leadership in world civilization is inseparably linked with climate and that with advance in culture it has been transferred toward colder lands, and when extant culture has declined, leadership usually has retreated southward . . . that the part of a civilization's banner has led steadily northward while culture was advancing and *vice-versa.*" Beginning with Montesquieu, and restated many times thereafter, the theory of equatorial drift was that peoples living at ease in the warm lowlands have been overrun by hardier races bred in the more rigorous climates further north or at higher altitudes.

The strongest hypotheses about civilization and climate were those of Ellsworth Huntington. He argued that climate is a decisive factor in health; that it determines physical and mental efficiency; that climate continually changes in time. He therefore concluded that climate determines the growth and decay of civilizations, their distribution on earth, and the historical destinies of nations. Since a civilization is the result of the energy, efficiency, intelligence, and genius of the population, and since these are determined by climate, therefore climate is the factor in the progress or regress of civilizations.

III. CONCEPTS OF DIFFUSION: EMERGENCE OF A NEW GEOGRAPHIC PARADIGM OF PROGRESS

As the environmentalist paradigm came to dominate American geography, alternative ideas were advanced and caveats offered; and it was out of these that both the ecological "reintegration" of man and nature and a new paradigm of progress emerged. Even that most ardent determinist, Ellen Churchill Semple, said that "It does not suffice that a people, in order to progress, should extend and multiply only its local relations to the land. This would eventuate in arrested development. . . . The ideal basis of progress is the expansion of the world relations of people, the extension of its field of activity and sphere of influence far beyond the limits of its own territory, by which it exchanges commodities and ideas . . . that intellectual force which feeds upon the nutritious food of wide comparisons. Every wide movement which has widened the geographical outlook of people . . . has applied an intellectual and economic stimulus" (*Influences,* p. 69).

Already, too, Franz Boas had begun to articulate the concept of cultural diffusion as a reaction against the unilinear evolutionism of E. B. Taylor and Lewis Henry Morgan, following from Ratzel's suggestive attempts to trace historical connections between cultural similarities. Boas was a German geographer whose role in the early development of anthropology in America is well known. He argued that "there is not a single people . . . that has developed its culture independently" (Boas, 1943).

Similar ideas are found in Vidal de la Blache, who distinguished between rudimentary cultures, the expression of local environments, and superior civilizations that can communicate with each other and are therefore capable of expansion. He equated stagnation with isolation that has "dried up the sap of invention," and evolution with innovations brought by contacts from favored nodal points to indigenous modes of life with sufficient vitality to use them. It is in the nodal regions that progress, he argued, has seldom been interrupted.

It remained for Carl O. Sauer to crystallize the essential ideas, however (Sauer, 1952):

> The focused curiosity that bears the name 'geography' is or should be aware not only of the dependence of life on the physical environment, but also on the interdependence of living things in a common habitat, or of total ecology. (p. 1)
>
> That succession of events with which we deal is quite other than the conceptual models that are set up as regular, recurrent, or parallel stages and cycles. . . . Among geographers, William Morris Davis delayed somewhat our learning about the physical earth by his systems of attractive but unreal cycles of erosion, with their stages of youth, maturity, old age, and rejuvenation. . . . Such concepts are sometimes, but improperly, called "evolutionistic." (p. 2)
>
> There is no general law of progress that all mankind follows; there are no general successions of learning, no stages of culture, through which all people tend to pass. There have been progressive cultures and others that show almost no

> signs of change. The latter are found in areas of high isolation; the former have been favored by the nature and location of their homelands . . . in a few . . . physically favored areas some . . . center has burst forth into a great period of significant invention, from which ideas spread, and in part changed as they spread afield.
>
> In the history of man . . . diffusion of ideas from a few hearths has been the rule; independent, parallel invention the exception. (p 3)

Precondition: Rejection of Evolutionary Frameworks

Environmentalist concern for the relations of nature and culture had given American geography its first explicit rationale, its first paradigm. The influence of environmentalism waned with a growing volume of criticism in the 1920s and 1930s (Tatham, 1951), however. Its principal failure had been that it tried to describe nature and culture as separate entities with direct lines of causation from former to latter.

At first, as environmentalism waned, signs emerged that the discipline might move on to William Morris Davis's later genetic-morphological frameworks, in which analyses were undertaken of parallel "natural landscapes" and "cultural landscapes" that developed from youth through maturity to old age and rejuvenation. But as the quotation from Sauer indicates, this framework also proved objectionable to geographers and, instead, in the 1930s and 1940s, American geography moved most affirmatively not along historical-genetic lines but along generic-functional paths in studying "areal differentiation" (Hartshorne, 1939) and, later, "areal functional organization." At best, relations between nature and culture were addressed in a framework of "possibilism," and only very recently has there been formal reintegration in most sophisticated philosophies of man-land relations, which see nature and culture as interlocking components of larger ecological systems (Wagner, 1960).

Meanwhile, distant from the mainstream of prewar geography, Sauer patiently studied the origins and dispersals of agriculture. His students extended the work to a variety of other phenomena—for example, the diffusion of grid-pattern towns (Stanislawski, 1946), and along the way waged a continuous and ultimately successful battle against the indiscriminate application by geographers of pseudoevolutionary frameworks, such as those that were borrowed from Turner's frontier hypothesis (Gulley, 1959) or articulated in the idea of sequent occupance, described in the preceding essay by Mikesell.

The Diffusion Model

The significance of the diffusion idea was that it moved away both from the place-bound frameworks of environmentalism and from the constraints of evolutionary thinking, each of which talked about relationships among things *at* particular places or within particular regions, thereby viewing change as *endogenous.* Instead, it looked at connections among areas, and *exogenous* sources of change. Thus, "growth does not appear universally at any one time but manifests itself at points or poles of growth . . . and diffuses through the economy in definite channels" (Perroux, 1955). Growth cannot, in this view, be separated from change produced by the diffusion of innovations, a process involving the "acceptance over time of some specific idea or practice by individuals, groups or other adopting units linked to specific channels of communication, to a social structure, and to a given system of values or culture" (Katz, Levin, and Hamilton, 1963). Moreover, the effect of diffusion is to continuously transform static man-land relationships, for "every time a new implement or machine is invented or a new technique is devised, a new appraisal must be made of every scrap of territory and the possibility of a new orientation of human activities be predicated. Areal distributions are essentially impermanent" (Kimble, 1951).

Beginning with Graebner's work on the theory of diffusion, Sauer's activities were an extension of the research on diffusion processes by American anthropologists such as A. L. Kroeber. The principal conclusion of this work was that "diffusion has always had a catalytic function in sociocultural development. The comparatively rapid growth of human culture as a whole has been due to the ability of all societies to borrow elements from other cultures and to incorporate them into their own . . . the necessity of integrating newly acquired elements into one's cultural heritage creates new problems, which demand new solutions and thereby engender new ideas. It was the opportunity for relatively rapid interchange of inventions and ideas between a number of local cultures that made possible the birth of the oldest civilizations in the Near East. It was stimuli emanating from these oldest civilizations which started the chain reaction that eventually resulted in the emergence of one civilization after the other through the whole of human history. Obviously, the cultures of marginal peoples least exposed to diffusion . . . have remained the most primitive known in modern times" (Heine-Geldern, 1968).

It remained for a Swedish geographer, Torsten Hägerstrand, to develop the first formal geographic model of diffusion processes (Hägerstrand, 1953, 1967). Hägerstrand looked at a society as an ordered system in which all individuals and all pieces of material equipment, including land, are component parts, linked together in a multitude of ways. Diffusion, for him, involved the introduction of some new element that could propagate itself until it permeated the system, and to some degree, transformed it. Certain macroconcepts, linked together, comprised his model. These macroconcepts summarize the recurring characteristics of diffusion processes:

1. *The growth curve.* Diffusion tends to be characterized by an S-shaped curve of cumulative growth, with slow take-off, rapid intermediate growth, and declining later stages asymptotically approaching a final ceiling. The absolute pace of diffusion varies widely (the "diffusion time"), but the curve is remarkably repetitive.

2. *Spatial expression of growth.* The three stages of the growth curve have recurring counterparts in the development of the spatial pattern—

(a) Initial stage. Adoption tends to be concentrated in a cluster of clusters of *centers of innovation.*

(b) Intermediate stage. Expansion takes place by means of a *neighborhood effect* in which spread is more likely closer to existing adoptions. Occasionally *news centers* of dispersal may arise from jumps of unexpected length.

(c) Saturation stage. The original areas of dispersal fill, while the frontier is still advancing, up to *barriers* and at a pace controlled by local *receptivity* to change.

Hägerstrand's most important contribution was to develop a model of the neighborhood effect, given initial centers of innovation, by translating the distance-decay pattern of personal contacts into a contact-probability grid (the "mean information field") that could be used to simulate spread, using Monte Carlo methods. Later versions of his model built in individual psychological resistances to innovation and physical barriers to contact. He also realized that long-distance "jumps" were to be expected, and he realized that urban innovations, in particular, tend to move from large cities to small down the urban hierarchy.

After Hägerstrand's contribution, work multiplied and finally achieved the status of an alternative geographic paradigm of progress. The general ideas in geographic diffusion theory today are those that relate to: 1. *Centers of innovation and spread.* Some centers, over long periods of time, emit wave after wave of innovations. Thus, Sauer tried to codify historical centers of agricultural innovation (1952), and Perloff et al. (1950) have concluded that, in studying American regional development, the central driving force has been

> . . . a great heartland nucleation of industry and the national market, the focus of large-scale national-serving industry, the seedbed of new industry responding to the dynamic structure of national final demand, and the center of high levels of per capita income . . .

and that standing in a dependent relationship to the hinterland,

> . . . radiating out across the national landscape are . . . resource-dominant regional hinterlands specializing in the production of resource and intermediate outputs for which the heartland reaches out to satisfy the input requirements of its great manufacturing plants. Here in the hinterlands, resource-endowment is a critical determinant of the particular cumulative advantage of the region and hence its growth potential.

In Perloff's view, the heartland experiences cumulative urban-industrial specialization, while each of the hinterlands finds its comparative advantage based on narrow and intensive specialization in a few resource subsectors, only diversifying when the extent of specialization enables the hinterland region to pass through that threshold scale of market necessary to support profitable local enterprise. Flows of raw materials inward and finished products outward articulate the whole (Ullman, 1957).

This view focuses on the particular role of metropolitan centers. It is they that organize the space-economy. They are the centers of activity and of innovation, focal points of the transport and communications networks, locations of superior accessibility at which firms can most easily reap scale economies and at which industrial complexes can obtain the economies of localization and urbanization. It is they, therefore, that encourage labor specialization, areal specialization in productive activities, and efficiency in the provision of services. Metropolitan centers, standing at the top of the nation's urban hierachy and at the center of each of its regions, are therefore the nodes whose connections integrate the national economy. Agricultural enterprise is more efficient in the vicinity of cities. The more prosperous commercialized agricultures encircle the major cities, whereas the peripheries of the great urban regions are characterized by backward lower-income economic systems.

The polarization of innovation in larger metropolitan areas is a reflection of trends in the national space economy. Differential growth rates among cities are a function of competitive advantages, but as Thompson (1968) points out, the long-run viability of a metropolitan area resides not so much in regional resources, traditionally viewed, but on its own capacity to invent and innovate, or otherwise acquire new resources and export bases. To follow his argument, he says that the economic base of the larger metropolis is the creativity of its universities and research parks, the sophistication of its engineering firms and financial institutions, the persuasiveness of its public relations and advertising agencies, the flexibility of its transportation networks and utility systems, and all the dimensions of infrastructure that facilitate the quick and orderly transfer from old dying bases to new growing ones. Larger urban areas, he argues, combine a favorable industry mix for growth with a steadily declining share of the various growth industries. High wage rates of the innovating area, quite consonant with the high skills needed at the beginning of the learning process, become excessive when skill requirements decline, and the industry (or parts of it) will then filter down to smaller, less industrially sophisticated areas where the cheaper labor can meet the declining skills demands of the filtering industry, thus creating the phenomenon of small towns with low-wage, slow-growth, filtered-down industry at the time when the metropolis has moved on to new bases. This capacity of the large metropolis to invent new economic bases means that large urban regions are free of narrow export dependency. At sufficient scale, infrastructure and residentiary development make the growth of the large metropolitan complex essentially self-generative, with a tendency to grow at or about the rate of the nation (Thompson, 1968). Moreover, the "circular and cumulative causation" of innovation breeding further innovation (Myrdal, 1957; Pred, 1966) continually reaffirms the growth of the metropolitan cores, and leads Friedmann and Miller (1965)

> . . . to interpret the spatial structure of the United States in ways that will emphasize a pattern consisting of *one,* metropolitan areas and *two,* the inter-metropolitan periphery. Except for thinly populated parts of the American interior, an inter-metropolitan periphery includes all the areas that intervene among metropolitan regions and that are, as it were, the reverse image of the trend towards large-scale concentrated settlement that has persisted in this country for over half a century. Like a devil's mirror, much of the periphery has developed a socioeconomic profile that perversely reflects the very opposite of metropolitan virility.

2. *Channels of spread.* Hägerstrand emphasized that, as innovations diffuse, urban centers tend to function as nodes channeling spread, but that over a broader range of innovations it is important to understand the role of the regular paths by which information is exchanged, including the spatial range of contacts displaying, most frequently, the distance-decay pattern (Bowers, 1937; McVoy, 1940) that leads to the neighborhood effect. Combining the two ideas, some have argued elsewhere that there are two major elements in the city-centered organization of economic activities in space that are important in molding the growth processes currently operating in the country:

> (a) A system of cities, arranged in a hierarchy according to the functions performed by each.
>
> (b) Corresponding areas of urban influence surrounding each of the cities in the system.

We know the following about this system of spatial organization:

> (a) The size and functions of a central city, the size of its urban field, and the spatial extent of development radiating outwards from it are proportional.
>
> (b) Impulses of economic change are transmitted in order from higher to lower centers in the urban hierachy, so that continued innovation in large cities remains critical for extension of growth over the complete economic system.
>
> (c) The spatial incidence of economic growth is a function of distance from the central city. Troughs of economic backwardness lie in the most inaccessible areas along the peripheries between the least accessible lower-level centers in the hierarchy.
>
> (d) The growth potential of an area situated along an axis between two cities is a function of the intensity of interaction between them.

To this, we should, of course, add the influence of the networks of communication and transportation connecting the areas of influence to the cities and the cities to each other, determining by their layout and efficiency the pattern and extent of spread effects, the depths of peripheral troughs, and the configuration of growth axes (Haggett and Chorley, 1969).

The recent diffusion of television illustrates these organizing themes (Berry, 1970c). Television stations were installed by cities in the familiar S-shaped logistic time sequence, although with gaps caused by two periods of war. The diffusion pattern of stations among cities was essentially hierarchical. Large cities installed stations before smaller cities and, holding size constant, heartland cities introduced television earlier than hinterland cities. The proportion of households purchasing television sets depended upon the hierarchical diffusion of broadcasting stations, but the spatial incidence of purchases was also a function of distance from broadcasting cities. As diffusion proceeded apace, market penetration neared completion first along the nation's principal growth axes and extended wavelike into the periphery. Continuing, total geographic coverage was achieved by 1958, and low degrees of market penetration thereafter remained only in the nation's troughs of economic backwardness.

Two contrasting views about growth processes that reappear in the literature intertwine here. One view (Friedmann, 1966) is that if metropolitan development is sustained at high levels, differences between center and periphery should be eliminated and the space-economy should be integrated by outward flows of growth impulses through the urban hierarchy and the inward migration of labor to cities. Troughs of economic backwardness at the intermetropolitan periphery should, thereby, be eroded, and each area should find itself within the influence fields of a variety of urban centers of a variety of sizes. Continued urban-industrial expansion in major central cities should lead to catalytic impacts on surrounding regions. Growth impulses and economic advancement should trickle down to smaller places and ultimately infuse dynamism into even the most tradition-bound peripheries.

This equilibrating theme is contested by Myrdal's notions of "unbalanced growth," that inequities tend to increase rather than decrease:

> In the centres of expansion increased demand will spur investment, which in its turn will increase incomes and demand and cause a second round of investment . . . the banking system . . . tends to become an instrument siphoning off savings from the poorer regions to the richer and more progressive ones where returns on capital are high and secure (Myrdal, 1957, p. 18) . . . (and which) are on the whole firmly settled in a pattern of continuing economic development, while (in the poorer regions) average progress is slower . . . in recent decades the economic inequalities between developed and underdeveloped . . . have been increasing. (p. 59)

If a "first-most" effect characterizes the beneficiaries of innovations, then it certainly seems that Myrdal's hypothesis should hold, because in most diffusion processes, the couplet is completed by "troughs last—and least."

3. *Boundary effects.* If there are regular centers of innovation and channels of diffusion, a third element, *boundaries* —physical, political, cultural—affects the pattern of spread by inhibiting contacts across the boundary. Such

boundaries may be completely impermeable, defining self-contained spatial systems, or they may be only partially absorbing in their effect, or even partially reflecting, so that innovations which would otherwise diffuse outside the system bounce back to some point within it. Such is the intent, for example, of tariff restrictions on trade.

4. *Receptivity factors.* Finally, differing groups and cultures may have differing capacities and willingness to adopt particular innovations, and thus may accelerate or retard the diffusion process.

IV. FROM SEGMENTATION TO INTERDEPENDENCE: CONTEMPORARY GEOGRAPHIC THEORIES OF SOCIAL CHANGE

Contemporary geographic theories of social change thus tend to focus on diffusion processes as the growth mechanism and to explore the ways in which innovations induce growth within, or lead to the transformation of, spatial systems. There is, for example, concern about the changes that take place in society when spatially-distinct locally-segmented societies begin to interact and to create new interdependencies, growing differentiation, and increasing complementarity. To cite one example, Soja's 1968 study of the geography of modernization in Kenya:

> . . . geographical patterns of modernization reflect the relative extent of social mobilization throughout the state-area and, by association, are closely linked to the development of a national metwork of social communications.
>
> The modernization society is . . . an emerging spatial system, a geographical community functionally organized to promote the interrelated processes of social, economic and political change. . . .
>
> Traditional society in Kenya, for example, was characterized by small units, ethnically circumscribed and inwardly focussed. Communications were almost entirely informal, personal and oral. Rarely did distinct circulatory organizations exist, for the flow of information was usually guided by social attributes, such as clan membership or position within a social hierarchy. The environment, especially the existence of physical barriers and the friction of distance, restricted both the extent and the intensity of communications so that the effective "world view" of any particular group was relatively narrow and usually confined to the group itself and its immediate neighbors.
>
> Colonialism had the effect of creating a new and stronger pattern of circulation within larger units of organization. With the drawing of boundaries and the spread of effective administration came the genesis of transition. Traditional society was virtually forced to change, and at a more rapid pace than ever before, through increasing contact with modern cultures.

An increasing ingredient of intentionality emerges from this view, for colonialism (itself an innovation) introduced revolutionary changes into the Kenyan spatial system, many by design. Most diffusion studies thus reemphasize the role of man as initiator, planner, executor, and achiever of goals.

Contemporary views of change have thus come full circle from Ritter's teleological "new geography" of the eighteenth century.

For much of his history, man and his geographers tended to see the content and ordering of his life as relatively fixed. Where change was or might be taking place, it was thought to be through the agencies of gods or fate, or the inexorable working of natural law or the forces of historical evolution. Man's role was limited to suffering the results, or cooperating with change to achieve latent, but strictly prescribed, potentialities.

Diffusion theory has played an important role in demythologizing nature and history and revealing man's role in change; thus geography now increasingly views man as a potential initiator: "expansion and modernization of transport communications is a key stimulus to growth in areas of the developing world where market place trading is a relatively recent innovation"; and "colonial use of taxing powers and creation of market places, together with road improvement, was instrumental in changing the economic and social development of the area" are but two of the relevant quotations culled from one recent work.

The view of man as an initiator spills over into the notion of the geographer as a planner. The realization of man's capacity to induce change in desired directions has thus led geographers most recently to new rounds of theorizing about the *spatial structuring* of growth to achieve stated ends. A new orientation has appeared in which goals are set, research carried on, and decisions made and implemented. To the extent that this process is successful, the accompanying change is intentional.

For example, from the understanding that the developmental role of urban centers in the United States involves the simultaneous *filtering* of growth-inducing innovations down the urban hierarchy, and the *spreading* of the benefits resulting from the growth both from core to hinterland regions and from metropolitan centers to the intermetropolitan periphery, geographers have been led to the formulation of growth center theory and strategies. One recent paper (Berry, 1970c) concludes thusly:

> Can growth centers be used to induce development in these lagging poverty regions (Parr, 1965; Hirschman, 1958)? If by "development" is meant ensuring a population growth rate approximately that of the nation, with steadily rising real incomes, the models presented earlier suggest several variables to be influences in achieving such goals:
>
> 1. *Threshold limitations.* Any policy that reduces thresholds will induce development to penetrate further down the hierarchy and out into the intermetropolitan periphery.
>
> 2. *Diffusion times.* Since the income effect is a declining function of time, any hastening of the diffusion process will bring a greater income effect to smaller and more distant areas.
>
> 3. *Accessibility.* Particularly in the case of house-

hold innovations, the extent to which families make use of new innovations in urban centers that have adopted them is a function of their access to these centers. Any decrease in effective distance will have a multiplicative effect on use of innovations, because the distance-decay effect is negative exponential.

The distinction is already drawn, therefore, in the sense of Amitai Etzioni, between diffusion in an ongoing society, which, in the absence of adequate collective social control, causes driftlike change of an incremental kind and occasional revolutionary transformations, and a self-transforming society which can exercise control over the forces of change. The growth center strategy has all three features required by any self-transforming society:

(1) It is *cybernetic* in that it involves the collection, processing (coming to conclusions), and use (introduction into the decision-making and implementation process) of knowledge.

(2) It has a *power-mobilizing* characteristic in that it implies social capacity to translate knowledge into policy and to redistribute the assets of society in the interests of collective action.

(3) It attempts to build *consensus* by bringing growth and redistributing wealth to what would otherwise be an increasingly dependent and alienated periphery.

Much recent theorizing in geography tends to emphasize the basic building blocks required for effective structuring of spatial systems: efficient locations of nodes and centers and their arrangements in hierarchies; efficient development of transport and communication networks linking the hierarchies of nodes; efficient movements and interactions over the networks; and the resulting surfaces and regions that flesh out the spatial organization. New books provide the details (Haggett, 1965; Haggett and Chorley, 1969; Berry and Horton, 1970). Network development is seen as playing three vital roles in spatial evolution, for example: generating *node-connecting sequences, space-filling sequences,* and *space-partitioning sequences* (Haggett and Chorley, p. 261). Regionalization is seen as a tool to achieve explicit ends (Berry and Wróbel). Theories of the urban hierarchy, of industrial location, of network structures and flows, of spatial interdependence, converge in the arena of planned change.

But the nagging doubts of environmental limits have reappeared. If we accept that man and nature are interlocking components of larger ecological systems, are there limits to the spatial structuring that man may undertake, which, if exceeded, result in societal degeneration? This antiphonal, but classical, theme is likely to engender yet another cycle of geographical theorizing about social change in the years to come.

3

THE FORGOTTEN HERITAGE: CHRISTALLER'S ANTECEDENTS

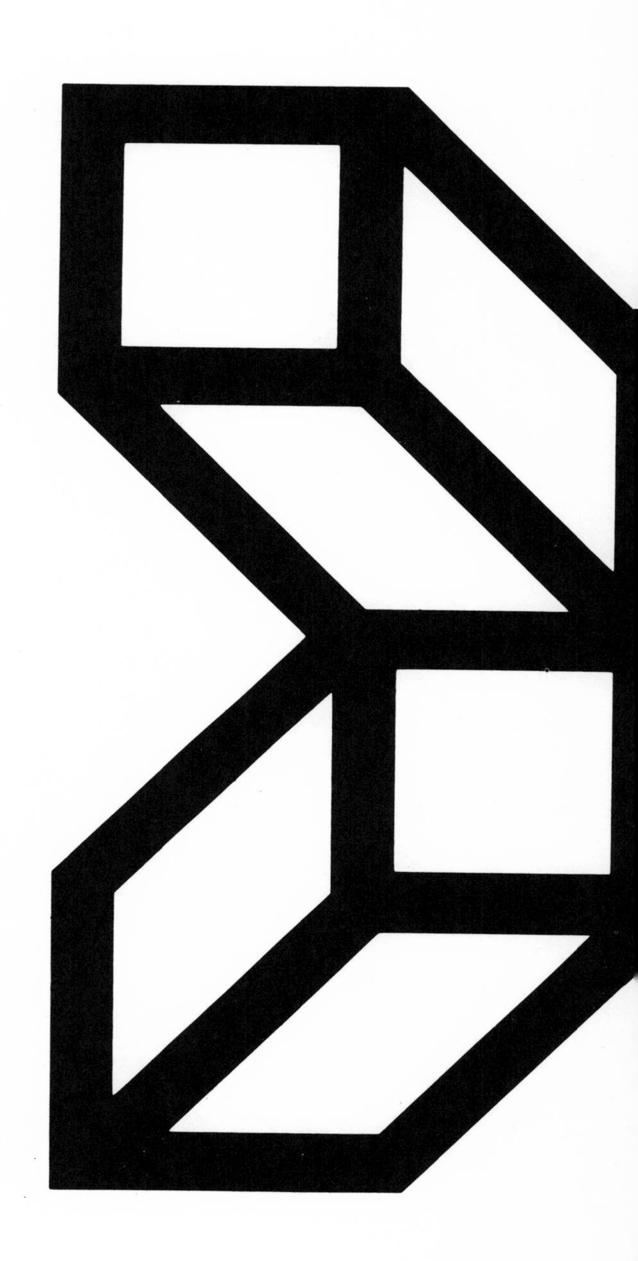

Christopher F. Müller-Wille
University of Chicago

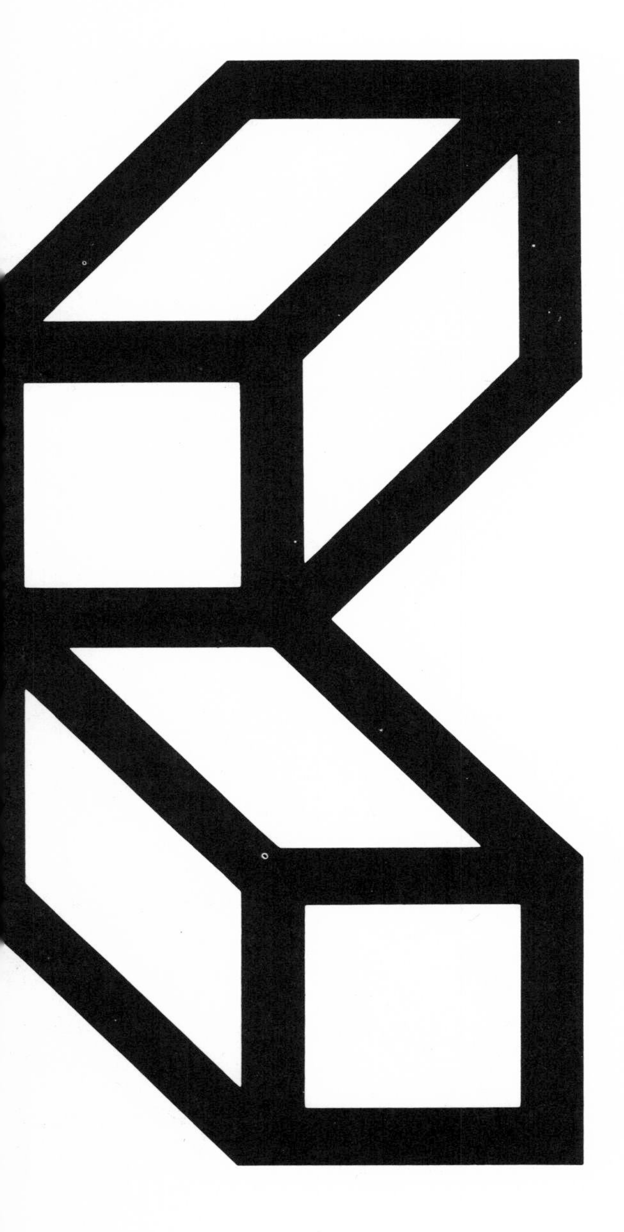

THE FORGOTTEN HERITAGE: CHRISTALLER'S ANTECEDENTS

Contemporary urban geography is a well-established and dominant subdivision of the field; yet it would be misleading to conclude that this fortunate situation is the result of a relatively long process of accretion and gradual expansion. As one examines the historical development of urban geography, it becomes evident that it was the introduction and eventual adoption of central place theory which represented the decisive impulse leading to the establishment of urban geography in its present form. The purpose of this essay is to offer an interpretation of the theory's relation to the structure and organization of early urban geography and to elucidate its conceptual contiguity to already-established research programs. The discussion focuses on two major topics. First, a number of studies which are believed to have laid the foundation of the field are examined. Second, representative examples of exploratory research activities conducted by urban geographers who were to become the geographic antecedents of Christaller are reviewed.

Such a retrospective view should help in the definition of major differences in conceptual frameworks and applied methodologies as well as in the identification of conformities regarding the views held before and developed by Christaller, thus contributing to the establishment of some standard measure by which to scale the theory's innovative character vis-à-vis previous paradigms of the field.

I. EXCURSUS: A CASE FOR SCIENTIFIC HISTORY

Any reasonably complete discussion which is formulated in explanatory terms as a self-contained system of statements—that is, a theory—will inevitably confront the problem of whether or not to include ideas and concepts which preceded its conceptualization and final articulation. Concomitantly, the stereotype question will arise as to whether such historical recounts provide plausible grounds for an exhaustive explanation of a newly established theory. A review of related antecedents rarely succeeds in accumulating sufficient critical evidence which promises to render satisfactory proof of the general need for a new, explanatory framework.

Such discussions seldom discover a steady, uninterrupted development of scientific understanding. Thus, Kuhn suggested that scientific growth cannot be reconstructed rationally. Progress in scientific knowledge takes place solely as a new general framework of explanation, or paradigm, is introduced and finds broad acceptance and application in scientific research and practice. As long as a theory allows the scientist to predict the results of his observations, it gives him all the understanding he needs. Hence, to believe in the continuity of scientific growth remains a conjectural consideration unless ongoing modifications or complete dismissal of a paradigm (a scientific revolution) are interpreted as discrete interruptions in the constant pursuit of normal science.

Normal science, by definition, furnishes a general research program which contains a concrete collection of operational concepts and defined methods: the intellectual and practical tools of any problem-solving activity. The execution of such programs, however, is not always determined by a series of rigorous procedural rules. Both the repeated testing of a paradigm's validity and the exploratory examination of yet excluded, marginal problems still constitute vital prerequisites to the continuing search for more comprehensive understanding. In fact, it would appear that the struggle for alternate explanations alone is conducive to increased knowledge. It follows that a recapitualtion of previous concepts serves the purpose of defining the circumstances under which, and not because of which, a novel paradigm is introduced.

In the following discussion the status and organization of urban geography before 1930 is analyzed on the basis of the above considerations. For obvious reasons, it proved impossible to prepare a comprehensive inventory of the discipline's evolution. Thus, emphasis is given to a precise identification and succinct examination of the most prominent concepts and prevalent modes of investigation commonly accepted and advocated by urban geographers at that time.

II. EARLY URBAN GEOGRAPHY: A STRUGGLE FOR IDENTITY

The remarkably resourceful study of rural and urban settlements in the then existing Kingdom of Württemberg, completed by Gradmann in 1913, represents a lucid illustration of both the internal organization of urban geography and its general status in human geography at that time. It was some seventy years previously, in 1841, that Kohl had published his innovative investigation of how varying geomorphic conditions, defined by their similarity to geometric forms and arrangements, determine the flow and direction of traffic and, thus, the location, growth, and spatial distribution of urban settlements.

Unfortunately, the following generations of geographers showed little interest in either expanding or substantiating Kohl's theoretical deductions. Moreover, they had overlooked the theory's potential use as the organizing concept for specific, empirical studies. Although the number of investigative field studies had increased at a steady rate, only a few serious attempts had been made to organize the study of urban settlements and formulate a coherent conceptual framework (Ratzel, 1891; Hettner, 1895; Richthofen, 1908). As a result, the field of urban geography offered an amorphous collection of incoherent geographic knowledge which was characterized by a persistent lack of both sound conceptual organization and comprehensive methodological guidelines. Commenting on this rather unsatisfactory situation, Gradmann candidly stated that

> the goals and methods of current research in settlement geography are still insufficiently clarified. We are particularly far away from having available universally applicable methods. One . . . who takes his research seriously, is consequently forced to develop methods on his own. (1913a, p. 5)

Having simply paraphrased a repeatedly voiced complaint, Gradmann accentuated a well-publicized dilemma of urban geographic research: the small community of interested geographers had been unable to institute the study of urban settlements as an independent yet integral element of geographic research. What were the underlying causes for this apparent failure? What chances were there of developing alternate routes of research? A detailed discussion of the many factors involved is beyond the scope of this essay. It seems worthwile, however, to touch upon some of the principal reasons that created the initial difficulties urban geographers had to overcome in identifying their specific domain of research so that some degree of communality could be established.

The Emergence of Human Geography

Early urban geographers frequently expressed the necessity of a general, conceptual framework which was to provide the basis for more independent and systematic research work dealing with human activities, their material manifestations, and temporal distribution over space. The most comprehensive effort to clarify the particular objectives of such a research program was made by Ratzel with his monumental, two-volume work *Anthropogeographie* (The Geography of Man, 1882). Ratzel's basic plan was to define the human element in geographic research and bring its evaluation in closer correspondence with the recognized system of geographic thought and explanation. He proposed a catalog of pertinent problems which geographers would be able to address, knowing that their solutions would be consistent with the principal goals of geographic research. That the study of anthropogeography was not yet considered an essential part of geography, but rather the intellectual link between (physical) geography and history is clearly demonstrated by the subtitle of Ratzel's work, which reads "The Principles of Geography's Application to History" (1882).

Such a definition of the association of the two disciplines obviously determined the scope of anthropogeographic studies, including the methods to be applied. The subject matter was to investigate the causal relationships between the natural environment—categorized as *Lage,*

Raum, and *Boden* —and human activities and movements in a historical, evolutionary perspective. In order to attain any geographically relevant results such studies had to employ methods that had been instituted by physical geographers, so that anthropogeographic laws could be obtained—laws which determine the varying modes of man's dependency on environmental conditions. The definite recognition of such laws was believed to provide the fundamental information the geographer needed to analyze the spatio-temporal distribution of anthropogeographic phenomena, specifically their origin and subsequent spread—classified as *Beziehungslehre, Verbreitungslehre,* and *Bewegungslehre.* This working procedure, Ratzel argued, would assure that geography, despite expanding its area of interest, could justifiably retain its proclaimed status as a natural science.

The rationale behind this pronounced bias, prevalent in nineteenth-century geography, rested on two programmatic postulates which had been advocated by Ritter to be essential as far as geography's competitive stance as a relatively young academic discipline was concerned. According to Ritter, the foremost purpose of geographic research was to analyze scientifically a traditionally philosophical problem: What are the invariable principles underlying the causal relationships between nature and culture, geographic place and historic event (Ritter, 1833; 1862)?

Both elements were believed to constitute an inseparable entity which was governed by a series of universal or divine laws. Nature was considered the medium, man the recipient or passive element structuring his activities according to natural conditions. The tangible object, from a geographer's perspective, was to be the earth's surface or the landscape in its totality. The areal differentiation of the earth's surface or the diversity of individual landscapes reflected the multitude of possible variations in the ways nature influenced man's actions. By assessing the physiognomic differences of geographic areas, the geographer was finally prepared to identify the specifics of universal laws and, thus, attain a fundamental understanding of the divine order governing the entire world.

Ritter asserted that to achieve this final goal the researcher had to resort to a particular methodological framework which relied exclusively on the general guidelines developed in the natural or "exact" sciences. The basic agreement with Ritter's elemental considerations led Ratzel to justify his initial claim of anthropogeography being a discipline which in the end contributed to the growth of knowledge in the natural sciences. Thus he presented a circular argument of Kantian philosophy that science must presuppose a causal law, since science itself can exist only because there is such a law.

The above outlined postulates emphasized two consequences which, at that time, seemed of vital importance to a consistent and sound development of general geographic research. The investigation of the causal relationships between land and man as expressed by the specific character of the total landscape—postulate one—and the use of similar methods in the various subdivisions of geography—postulate two—were to create the desirable internal coherence of this young discipline and eventually provide external stability. Both were needed not only to define geography's area of interest but also to distinguish it effectively from subject matters claimed by other sciences, namely, history and geology.

The generally felt obligation to observe these rigid guidelines undoubtedly helped geographers to hurdle elementary conceptual as well as methodological difficulties and to find the necessary identity and outside recognition. But as research progressed, it became apparent that Ritter's assertions had in fact established a program that was too narrow in scope to allow exploratory research, conducted at the frontiers of geography and especially human geography, to be utilized. Ratzel seemed to have sensed the limitations of Ritter's concept, but, unfortunately, did not succeed in articulating his doubts.

The Site-Situation Syndrome

It is not surprising, then, that geographic studies which dealt with urban settlements analyzed the location, historical development, and spatial distribution of cities in terms of their dependency on environmental conditions. A distinction was made between local and regional physiogeographic determinants—that is, topographical vis-à-vis geographical conditions (site-situation). A prime example of this research orientation seems to be Hettner's (1895) discussion of the origin and development of urban geography and its contemporary objectives and methods. Referring to Ratzel's work, the author flatly declared that to be in accordance with conventional procedures urban geography had to adopt the specific methods of inquiry and interpretation successfully exercised by physical geographers, especially plant geographers. Only the thorough understanding of the areal variation of environmental conditions on both the local and the regional level permitted the geographer to classify urban settlements, compare them, and finally arrive at the formulation of generally valid laws explaining the location, size, and historic evolution of single cities and groups of cities.

The urban settlement was compared to an organism which, much like a plant, had to adjust to the natural environment and compete with others in order for it to succeed in its constant struggle for survival and supremacy. This meant that unless a city was situated at the most advantageous location, as defined by the sum of favorable physiogeographic determinants, it was bound to decline in size and importance, if not be abandoned. Thus the effects of human activities were considered a noise factor or error element temporarily disrupting the continuing process of man's adjustment to the intricate dynamics of the natural environment as the ultimate corrective.

With this physiological approach, Hettner supported the rationalist's argument that nature represents the absolute standard of order and purpose. This was clearly demonstrated by the strict application of the Darwinian concept of survival by competition and adjustment. This view constrasted sharply with Ritter's pronounced belief in a transcendental source of order. With the exception of this timely change in philosophical creed, Hettner fully complied with Ritter's principal notions concerning the aim and methods of geographic research. The basically congruent attitude found its expression in the following statement:

> There is no reason to declare a dualism, a dichotomy of geographic research into physical geography and the geography of man. Although geography incorporates the study of both nature and man, the discipline itself remains undivided as to its objectives and methods. (Hettner, 1895, p. 375)

It would appear from the above statement that human geographers in general and urban geographers in particular had resigned themselves to leave the instituted postulates unquestioned and settle within the narrow confines of conventional, if not predictable, research activity. Bound by the everlasting obligation to uphold the principles of geographic inquiry and explanation as developed by Ritter and explicated by Ratzel, the human geographer's quest for a more liberal and independent working basis seemed to have dissipated. Nevertheless, Ratzel's work in particular proved to have broken new ground for future geographic research. Although it lacked an innovative and consistent perspective, it revealed an abundance of ideas and suggestions which inspired geographers to expand the limits of geographic observation and interpretation.

III. THE "NEW" URBAN GEOGRAPHY: A STRUGGLE FOR ALTERNATIVES

As the interest in human activities and their diverse influences on the physiognomic structure of the natural environment increased, a growing number of geographers tended to disapprove the absolutism of Ritter's basic conditions which purported nothing better than geographic eclecticism: a dead end for the advancement of scientific understanding (Troll, 1947, pp. 23–24). In reviewing the more recent developments in human geography in general and assessing their significance to the study of urban settlements, Schlüter, for example, argued that

> it is a serious misconception to believe that . . . anthropogeography can deal with man's dependency on nature or nature's influence on man. . . . The objective of geographic research is to understand the form, shape, and spatial arrangement or association of geophenomena which are visible on the earth's surface and thus observable. . . . To solve this problem geography has to have complete freedom as far as the explanation of these phenomena is concerned. All kinds of influential factors have to be taken into consideration may they relate to either or both environmental conditions and/or man's free intellect. (1899, pp. 66–67)

This statement of Schlüter's signified the beginning of a new, promising era, during which human geography was to develop to an equally respected, if not dominant, subdivision of geographic research as compared to physical geography. It was during this period of change that urban geographers had the opportunity to formulate their objectives more precisely and articulate the outlines of a unified conceptual framework.

The Consolidation of Human Geography

The major concern of geographic investigation was now focused on the landscape as man's habitat. Human activities were considered to represent the most influential force in changing the morphology of the natural landscape, in fact transforming it irreversibly to a world of temporal purpose and human design: the cultural landscape. Consequently, geographers defined the cultural landscape as "an area made up of a distinct association of forms, both physical and cultural" (Sauer, 1925). Phrases like "the morphology of the cultural landscape" and "the development of the cultural out of the natural landscape" are characteristic symptoms of the progressing change in the geographer's evaluation of the scope of human activities (Schlüter, 1906a; Sauer, 1927). A twofold research orientation emerged: (1) the study of the surface manifestations of man's present occupation of the earth; and (2) the investigation of their evolution to the contemporary material expression.

In accordance with the classical concept of geography having not only a chorologic but also a chronological aspect, it was suggested that the primary goal of human geography was the reconstruction of the various stages in the development of the cultural landscape as it evolved from primitive beginnings to present-day complexity of form. Although this research direction is most intrinsically associated with historical geography, the underlying purpose of such historical investigations was to prepare a broader basis leading to a more comprehensive understanding of the modern cultural landscape, since it incorporates the multifarious manifestations of present human activities as well as relics of previous time periods.

Particular attention was given to the evaluation of the historical development of rural settlement complexes and their surrounding, agriculturally used areas (Meitzen, 1895; see also Uhlig, 1972). In this context geographers specifically analyzed the formative influence that defined rural units had on the development and increasing morphological differentiation of the cultural landscape. Their preoccupation with this problem induced a number of geographers to consider the terms rural and cultural landscape as synonyms. It was argued that primitive agricultural activities represented the first definite step sedentary man took to interfere with and permanently upset the natural order and balance of the natural landscape or *Urlandschaft.* Hence the setting of rural environment provided the ideal laboratory for human geographic field work which was designed to pinpoint the various transitional stages from the prehuman to the cultural landscape. The difficulties in defining the undisturbed natural landscape are apparent. They, however, did not preclude geographers from arriving at erroneous generalizations. A case in point is Gradmann's assertion that the initial occupance in Central Europe took place in the open regions of heath vegetation (*Steppenheidetheorie*) (Gradmann, 1901, 1937).

The Urban Landscape

In contrast to this burgeoning branch of human geography, urban geography seemed to have decreased temporarily in its relative importance. Urban geographers generally tended to ignore the broader implications of the above mentioned specialized studies. Researchers displayed little or no intention of assessing the city's role in the formation of the cultural landscape, although urban settlements were considered a symptomatic, if not a critical, element in this historic development. They furthermore were not inclined to evaluate systematically the growing significance of cities to the accelerating transformation of the structure and organization of the cultural landscape at the time, be it with regard to the emerging locational pattern of secondary and tertiary activities or the spatial reorganization of agricultural production. Instead, they directed their attention to individual cities, which were defined as being the basic areal unit of proper urban geographic field investigation. This rather selective identification of urban geography's object might to some extent explain the field's temporary isolation from the mainstream of geographic research.

The majority of urban geographers described the city as an individual, morphological entity separated from the surrounding rural area. The rudimentary definition of landscape was commonly accepted as the operational framework. This permitted the formation of the urban landscape or townscape (*Stadtlandschaft*) as the organizing concept of urban geography, temporarily cementing the classic rural-urban dichotomy (Passarge, 1930). The then generally recognized definition of the city reflected quite distinctly this particular perspective. Size, density, heterogeneity, and the unifocal orientation of urban characteristics—that is, morphologically distinguishable phenomena and statistically controllable data—were singled out as common denominators (Hassert, 1907).

According to these collective categories, urban geographers were able not only to characterize individual urban landscapes but also to distinguish them from the generic class of rural settlement units. Within these general groupings particular attributes were to be identified. They, in turn, could readily be observed and described in all detail with regard to either their spatial distribution within the city or their individual morphological configuration and association. It would appear, then, that the primary objective of current investigations in urban geography was the detailed description of the various structural elements of the urban landscape, which was frequently accompanied by a sequence of illustrative maps. The underlying reason for this emphasis on minute observation and visual reproduction was that form, shape, and spatial arrangement were believed to represent the genuine manifestation of both function and process, be they contemporary or historical, physical or societal in nature. Time lags in this relationship were considered insignificant as long as the definite dependency of material evidence on formative force could be established.

The Schlüter Model

The foundation for the general change in research perspectives, which resulted in the rejection of Ratzelian unitary interpretation and led to the consolidation of human geography, was laid by Schlüter's works (1899, 1906a, 1906b). In his programmatic article "Notes on Settlement Geography" (1899) he set out to reformulate the common basis of general geographic research and, furthermore, attempted to outline the relevant objectives of contemporary urban geography. According to him, it should not be the ultimate task of geographic investigations to focus on an analysis of the principal physical or historical factors which might singularly or in combination determine the particular interrelationships between form and function, structure and process. It was argued that such a working procedure far too frequently revealed an unjustified bias in perspective and interpretation. The dogmatic selection of information and explanation seemed to have been adopted for the sole purpose of reaffirming a priori conditions.

Schlüter proposed that the unifying viewpoint of geography had to rest on a careful description of factual forms on the earth's surface. Consequently, scientific scrutiny should rather be applied to the visible product of a yet unknown number of independent variables, and not to a selected number of presupposed, causal factors. In any case, the process of identifying the latter seemed to depart from premature conclusions, if not preconceived theory. Both of these conditions threatened to limit the researcher's choice of valid information considerably and to obscure his ability to come to a pertinent judgment on the degree of a variable's explanatory power. In concluding his introductory comments Schlüter again stressed the vital necessity of unprejudiced, inductive field work:

> The only profitable approach (to geographic problems) depends on the most accurate and methodical researching of facts. This alone can give the addressed problem its tangible form and represents the indispensable foundation for any theoretical construct. (1899, p. 75)

As for the practical application of these general guidelines to urban geographic studies, Schlüter pointed out that the physical structure of urban settlements had thus far not been subjected to extensive, critical analysis, whereas the location, historical development, and spatial distribution of cities had traditionally been the favored objects of geographic curiosity and explanation. These investigations proved to have fallen short of producing any reasonable and conclusive answer. The blind desire to detect the directive forces behind these processual events permitted the researcher to acknowledge only a limited, predetermined set of independent variables, thus indulging methodological-philosophical prejudices instead of emphasizing the necessity of substantive, empirical work.

Schlüter offered two alternate routes of practical research in urban geography: (1) the analysis of the areal extension of urban influence and its material expression; and (2) the description of the internal structural differentiation of cities. The ensuing discussion started with the convincing argument that the conventionally defined city constitutes in reality the center of a larger, urbanized region. With increasing distance from the "central city," its influence would diminish in a regular fashion—that is, the number of distinctly urban characteristics decreased, creating under ideal circumstances a graded sequence of concentric zones. Schlüter tentatively listed them as: (1) the proper city or "central city"; (2) the inner ring of urban influenced and oriented settlements; (3) the zone of commerical interdependence; (4) the outer ring of long-distance trade and traffic routes and the sphere of cultural dependency and institutional dominance. In general, it was suggested that any settlement, ranging in size from hamlet to metropolis, generates a sphere of influence which extends beyond its immediate communal boundaries. The range of influence is determined by a settlement's size and proximity to others. This basic notion of a rural-urban continuum was subsequently modified in so far as only those settlements which developed and maintained a complete system containing all four rings were classified as urban.

The distinction between these zones was based on settlement-morphological differences, principally referring to the apparent, systematic variations in the intensity of land use. Indices selected to describe the differences included the height of buildings and the ratio of built-up area to open, unused space. In this context Schlüter called for substantial improvements of the available, yet scant, census materials which would undoubtedly facilitate the otherwise laborious and time-consuming investigations and ultimately be instrumental in preparing an operational basis susceptible to more comprehensive explanations.

To define the areal extension of individual urbanized regions or unifocal, zonal systems with reasonable accuracy was to become the central objective of this specific research theme in urban geography. The most intriguing aspects, however, related to the evaluation of more complex, regional situations, specifically, city clusters and their potential impact on the spatial structure of several contacting or overlapping zonal systems. In what way did such locational constellations of centers influence the range of individual zones? What were the consequences in terms of city-size frequencies and a city's relative position in a sequence of regional subordination and dominance? Was it realistic to contemplate the emergence of multi-nodal, regional systems as a possible development? Unfortunately, Schlüter displayed no further interest in pursuing these questions; neither did he attempt to explicate the procedural structure of such a research program. Instead he afforded the greatest part of his attention to the minute description of urban morphology.

Schlüter interpreted the central city per se as representing the most appropriate object in urban geography. This allowed him to apply two standard procedures of geographic inquiry: (1) to investigate the historical development of a city, thus defining the factors of its growth and partially explaining its structural diversification; and (2) to discuss the contemporary distributional patterns of selected urban characteristics, individually or grouped. He insisted that the complex question of what agents or causal factors had been most dominant in the development of cities could be answered on an individual basis only. In accordance with the current zeitgeist, the influence of historical events, or more precisely the visible material consequences of human activity and decision-making, was thought to be more significant than physiogeographic conditions in explaining the physical character of cities. By relying mainly on historical accounts that reported on changes in a region's sociocultural, political, and economic situation, the researcher was equipped with the necessary information which let him determine some of the reasons for the diversified structure of present-day urban morphologies as manifested by the distinctive differences between urban districts in terms of their architectural individuality and geometry of street plans.

The primary objective of urban geography, however, was to remain the detailed description of the contemporary urban landscape. Schlüter proposed that such investigations were best organized according to three elemental categories which formed a substantial part of the conventional definition of cities. They were area, population, and buildings. The raw data available on any of these categories were frequently presented as size or density calculations. In this form they already provided essential information about the internal differentiation of individual urban landscapes. A more comprehensive approach aimed at the identification of descriptive attributes, which were commonly listed under three subgroups: land use, function of buildings, and population characteristics—that is, socioeconomic qualifications. Although such lengthy catalogs, if researched carefully, seemed to preclude extensive investigations of larger cities or groups of cities, such as in regional, comparative studies, Schlüter definitely encouraged this mode of research so as to establish a sound foundation for the classification of urban settlements. Figure 1 delineates the topical organization of urban geography as discussed by Schlüter. Broken lines indicate dependencies between variables whose validity had been questioned.

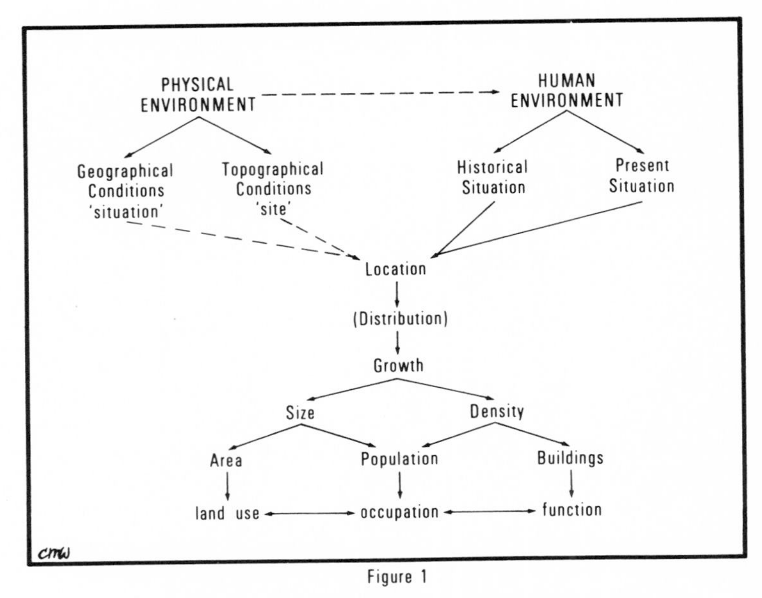

Figure 1

The topical organization of urban geography (after Schlüter)

Morphography vs. Morphogenesis

In the following decades the greater number of investigative studies dealt exclusively with the morphological differentiation of individual urban settlements. The major portions of such studies contained a painstaking description of the various distributional patterns of selected structural elements. Based on such meticulously researched accounts, three ideal-type urban morphologies were somewhat hesitantly suggested, most frequently with direct reference to the intraurban street plan and/or architectural differences. Each was believed to be the result of markedly different developmental stages in the historic growth of urban settlements. Schlüter's probing suggestion to consider such distributional patterns as being possibly generated or reinforced by contemporary forces calling for either the concentration or dispersion of intraurban forms and functions as determined by socioeconomic needs of daily or periodic character were completely ignored (1899, pp. 71–72). Yet, in another way, the distinction between these particular models appeared to be an overdue extension of a concept which was originally launched by Kohl in 1841 (figures 2a & 2b). Today, these distributional patterns are referred to as the concentric, semisectoral, and radial models (figure 3).

Unfortunately, further explicitly theoretical or plainly speculative analyses were dismissed as being nongeographical, if not unscientific altogether. Researchers continued to stress the unalterable necessity of inductive, empirical field investigations, thus attempting to prevent

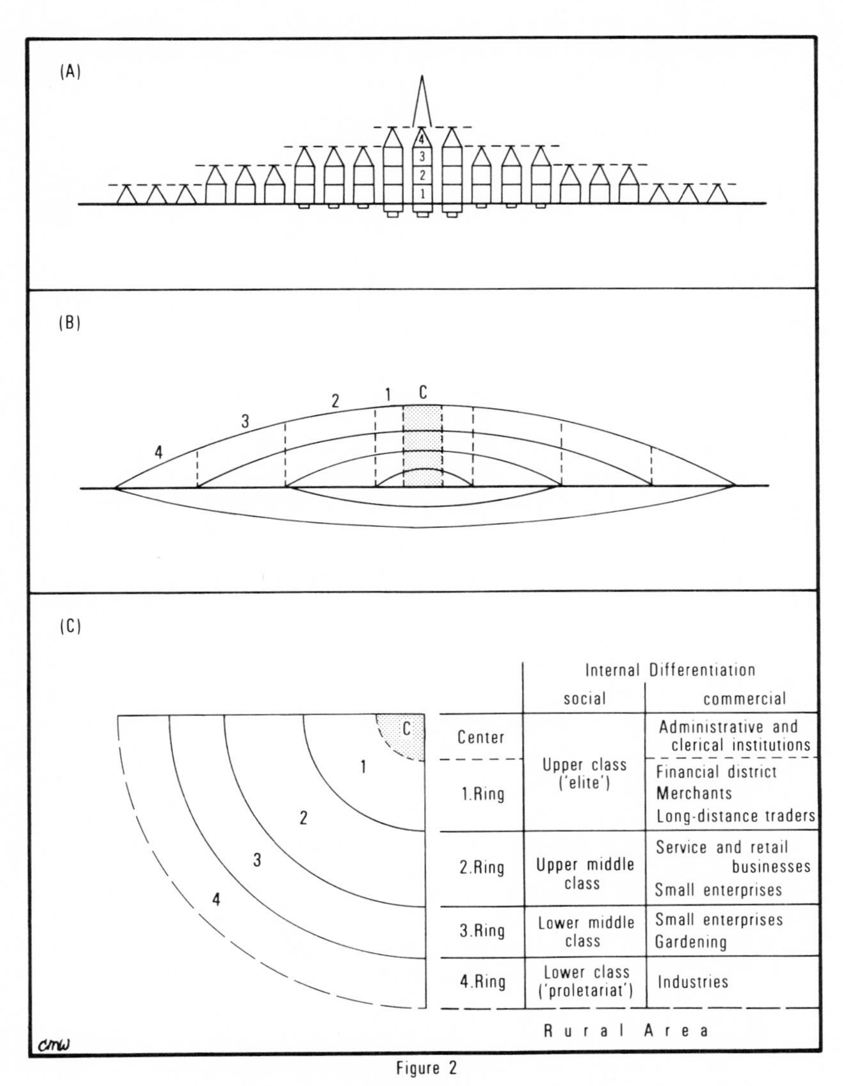

Figure 2

The ideal-type city: Its internal socio-economic and morphological differentiation (after Kohl)

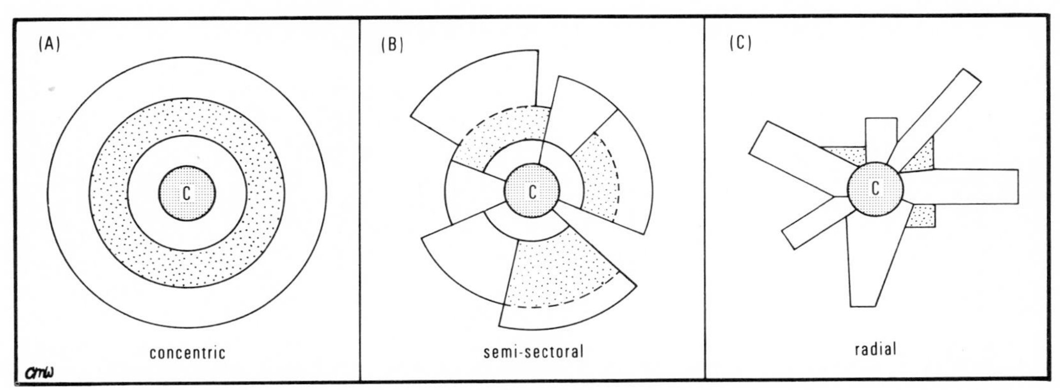

Figure 3

Ideal-type city morphologies

untenable generalizations or theoretical simplifications (Geisler, 1920; Carlberg, 1926). In order to attain a more intimate knowledge about the character of individual urban landscapes, it was considered necessary to identify the particular spatial patterns of so-called form-function correlates. As already indicated by Schlüter's principal discussion, this meant to dissect the physical structure of cities in terms of land use patterns and functions of individual buildings. To illustrate the basic working principle the reader is referred to Kohl's intuitive interpretation of the ideal-type locational association of urban characteristics (figure 2c). Here the morphological differentiation is compared to the extrapolated distribution of social classes and institutions as well as commercial enterprises.

According to these three categoric distributions, it is possible to establish a threefold relationship bundle. The first two links describe the organizational structure of the intraurban, socioeconomic space as being reflected by a distinctly graded sequence of morphological complexity—form-function correlates. The third comprises the inherent locational congruence of social stratification and economic activity—functional correlates. The latter of the three was explored by a negligible number of geographers, since the ultimate goal was to characterize the physical structure of cities and not their social diversity and economic base. The concept of form-function correlation, however, proved to be the conceptual backbone for any reasonable classificatory scheme structuring any regional or cross-cultural comparison (Geisler, 1920 and 1924; Passarge, 1930).

Since all these investigations dealt basically with the contemporary urban landscape and its internal morphological differentiation, they were commonly referred to as morphographic studies. A growing number of geographers, however, insisted that such relatively restricted investigations should be supplemented by detailed historical accounts. This procedure allowed the analysis of the various phases in the development of individual cities as well as groups of cities, thereby probably touching upon the long-standing problem of city-size frequencies and the spatial distribution of settlements per se. This mode of investigation was later on termed the morphogenetic approach. Related studies were conventionally divided into four segments: (1) situation or geographic situation; (2) site or topographic situation—both segments comprised an evaluation of the physio-geographic conditions as well as a discussion of the city's relative location in regional transfer systems; (3) historical development; and (4) morphology of the present urban landscape. Of the numerous works exploring this modified research direction, Gradmann's regional inventories of southwest Germany are the most prominent and comprehensive (1913b, 1916).

Supported by carefully researched documentary evidence, Gradmann came to the conclusion that the emergence and subsequent development of urban settlements, and hence their morphological structure, were primarily determined by politico-economic priorities of regional and national dimensions. Since urban settlements were defined on legalistic grounds rather than geographic evidence, it should be apparent that they were to be conceived as a distinct generic class of settlements. Thus they could be considered urban at the time of their foundation or legal designation. Planned, colonial, and imperial cities were the examples cited (1916).

It would appear that the morphogenetic approach, if applied in a regional or comparative context, furnished a plausible concept for investigating the somewhat neglected problem of identifying the principal factors which determine the location, historical growth, and spatial distribution of cities. Unfortunately, the principal arguments supplied by proponents of this view generally proved to be of limited validity. They applied only to specific regions or locations, and even then were severely qualified temporally. It was because of these apparent constraints, if nothing else, that geographers—who gave preference to the morphographic point of view—tended to consider these rather speculative discussions to be mere intellectual exercises having more or less propaedeutic character (Geisler, 1920; Carlberg, 1926).

In reviewing the status and organization of urban geography as of 1930, Dörries, for example, was unable to reconcile this controversy about what approach was more powerful in explaining inherently geographic problems. He was well aware of the general shortcomings of the research results thus far produced. His hope that the enormous wealth of still individualized information and tentative explanation might provide the requisite foundation for articulating explanatory generalizations was shared by many of his colleagues. In concluding his retrospect, Dörries made the following optimistic remark concerning the future objective of urban geography and geography in general:

> It is appropriate that up to now analytical studies have prevailed in geographic research. The more extensive and varied the collection of observed facts, gained by means of exact and detailed investigation, . . . the more certain will [geographers] be able to guard themselves against irrelevant research projects and erroneous conclusions. However, [they] should not think of summarizing the many findings with a more general perspective at too late a time, for only a synopsis guarantees the realization of essential relationships . . . As for future research activities geography, like any other discipline, cannot possibly afford to lack the vital collaboration of analysis and synthesis. (1930, p. 325)

A small number of researchers, however, had expressed their doubts as to whether this could seriously be considered a probable development. They asserted that the concept of the urban landscape, with its absolute emphasis on morphological elements, restricted the scope of investigation. A variety of urban characteristics, be they formal or functional, were still disregarded. Under these circumstances, it seems to have been a matter of time until urban geographers ventured to explore alternate routes of analysis and explanation, either by modifying the established research program or by proposing a radically novel approach.

The Economic Landscape

Since the rejection of Ritter's eclectic postulates and the institution of Schlüter's research program, there had been sporadic instances of further exploratory research activity that attempted to come to a more comprehensive understanding of cities, in general, and to identify the principal human actions influencing the growth of cities and their irregular spatial distribution, in particular. The basic question being asked was: What makes some people live in cities, and others in rural communities?

The investigations concerned with this problem, despite their occasional ambiguity, concurred on one principal argument, in that, without exception, they stressed the functional and spatial division of economic activities as the fundamental condition for the emergence of urban settlements. The elemental, rather descriptive distinction between natural and cultural production, or areal and point production/supply, functioned at first as conceptual component, so as to enforce the conventional definition of cities—that is, it underscored the concept of the rural-urban dichotomy (Hanslik, 1909). A relatively more elaborate approach was taken by those who examined two characteristic journey-to-work patterns, distinguishing between centripetal, or urban, and centrifugal, or rural movements. According to the first, as substantiated by daily commuting statistics, the sixty-minute isochrone was described to represent the maximum feasible distance for daily commuting to the central city (Hassinger, 1910a and b). It was suggested that this functional definition provided a more sensitive basis of identifying the outer limits of urban areas. Although this elemental problem had not remained unnoticed, the nonchalant unconcern shown by the majority of urban geographers is rather surprising (Hasse, 1891–1892; Schlüter, 1899; Blanchard, 1922). The rapid growth of urban centers had made it increasingly difficult to define the limits of the urban landscape. Urban morphologists especially depended on a precise definition of their study area. Their failure to establish appropriate and consistent indexes made such definitions improbable and, furthermore, prevented the successful incorporation of such processes as urbanization and industrialization. These constraints again emphasize the conceptual as well as operational limitations of both the morphographic and morphogenetic approach.

Economic Competition

A more specific analysis, which was designed to elucidate the fundamental processes linking urban economic activities with locational and physiognomic characteristics of urban settlements, was advanced by Hettner (1902). He had eventually modified his formerly conservative stance concerning the proper methodology of inquiry and explanation in urban geography (q.v. 1895, n. 6b). In his latest, rather brief discussion Hettner departed from the traditional interpretation, holding that the spatial organization of transfer networks determines, as an indirect factor, the location of cities as well as their economic functions (Ratzel, 1903). The leading arguments of Hettner's excursion into the fields of hypotheses and deductive logic were derived from the assumption that urban settlements emerge and develop at those locations that guarantee the profitable operation of specialized economic activities. Hence the author contended that a careful classification of urban settlements according to their economic differentiation would yield the information essential to identify the principal variations in the economic structure of cities and their particular effects on the location, size, morphology, and spatio-temporal distribution of cities.

It was argued that the inherent regularities of economic competition—in other words, the need for sufficiently large market areas—result in a number of space-organizational regularities in the distributional pattern of business locations. First of all, the economic composition of cities will determine their size and physiognomic differentiation. Secondly, and more significantly, the mix of economic functions offered some definite clues as to what kinds of relationships might develop between cities of similar and/or different economic organization.

With reference to the former, it was suggested that cities of comparable economic structure are spaced in a regular fashion, since each one needed to control a market of similar size, so as to maintain its economic viability—territorial exclusivity. In more general terms this meant that cities of equal economic character tend to be situated at locations which offer similar space-organizational conditions. In this context Hettner called for the measuring of average distances between cities of equal size. As regards the latter category of interurban relationships, one might detect that urban settlements with small market areas are subordinate to those with larger ones. In both cases the growth of settlements was interpreted as being the direct result of the continuous struggle to enlarge one's area of economic control. The proximity to other cities and the degree of similarity in terms of economic potential would finally set the relative limits of physical and economic growth.

The shift in Hettner's logic had been rather subtle. In a way, it seems to have consisted of a mere change in wording. Analogous to the original supposition of biological competition, Hettner now proposed to consider the drive for economic profitability and sustenance as the fundamental forces determining the economic structure of cities and the spatial distribution of classified groups thereof. Hettner's theoretical reflections bear an astonishing resemblance to Christaller's principal course of argumentation. Considering the initial reception of the central-place model, it should not be of any surprise that Hettner's rudimentary concept of a space-economy was brushed aside. The major reason for this negative response seems to have been the widespread realization that proper geographic research was to bear on individual groups of observed phenomena, and not so much on their broader connections. In general, geographers felt compelled to exclude any speculation upon the ultimate causes or origins of facts or phenomena. Consequently, the entire discipline underwent the definite conversion toward empirical studies of specific details. It was commonly argued that carefully researched field examinations alone could fathom the complex systems of interdependencies. Such detailed studies would then generate the essential clarity, allowing them to come to a true understanding of reality—that is, to develop an analytical knowledge of the temporal organization of human space and time, thus forestalling the inherent possibility to be entrapped by the obscurity of futile assumption and premature generalization.

Local Exchange Markets

Gradmann, in compliance with this emerging theme of practical geographic research, set out to gauge the extent to which local and regional economic conditions exert a directive influence on the development of individual urban settlements (1916). Supported by an intimate knowledge of the historical development of cities in the study area, Gradmann concluded that two principal processes of urban growth could be distinguished: (1) the primary formation of self-contained systems of local exchange markets, and (2) the secondary development of linear, interurban trade and regional traffic networks.

As to the first, it was suggested that the immediate rural surroundings of cities create the primary conditions not only for the emergence of cities but for their initial development as well. The urban settlement was described as containing functions aside from those that rendered services to its own population. Its primary function was to provide a convenient, accessible center for local trade and traffic. Thus the urban settlement operated as the central marketplace, monitoring the exchange of local goods produced in the surrounding area (umland). In addition it offered specialized services for the rural population. According to these elemental functions, the initial size of an urban center depended primarily on the degree of interaction between rural umland and the center. The intensity of this process was determined by the size or, more specifically, the productivity of the rural umland, and by the number of specialized urban services. The result of this interplay would be the emergence of numerous small- and medium-sized cities. Urban center and rural umland formed a self-contained economic exchange system.

Gradmann suggested that additional growth is induced by the development of long-distance or interurban trade and traffic. On this regional level the local center would have to function both as a port for locally produced goods to be shipped to other markets and as a distributor of imported products either to its local rural and urban population and/or to other cities. To illustrate the effects of this secondary process of additional urban growth, Gradmann described the impact of regional trade unions on the development of individual urban settlements. The institution of a regulated, regional transfer network induced integrated cities, because of the addition of new functions, to grow beyond the size which could originally have been supported by the rural population. As the regional interurban trade activities declined or were disrupted for a longer period of time, formerly participating cities would dwindle to their normal size and importance, as determined by the economic potential of their surrounding rural areas.

The interaction of the two processes was interpreted to be the major cause for the empirically established, irregular distributional frequencies of city sizes. It was surmised that the two formative processes were characterized by considerable variations with regard to their spatio-temporal stability. In general, it would appear that Gradmann considered the emergence of long-distance interurban trade as a perturbating element offsetting the inherent space-economic equilibrium of local market systems. Thus he strongly contested the traditional assumption that long-distance trade routes and their operational characteristics in particular (for example mechanical and commerical breakpoints) constituted the primary incentive to the foundation, location, and initial growth of urban settlements.

The most significant result of Gradmann's inductions appears to have been the definite recognition of the indivisible territorial and functional areal unit created by the elemental economic interdependencies between urban center and rural umland. Such integrated, local systems of human interaction—in this case specified by the mechanisms of economic exchange—were believed to represent the original cells of urban development. By accepting the notion of the city being the center of an abstract, space-organizational entity, it was possible for the alert worker to develop a more sophisticated, analytical approach. It happened to be Bobek who finally succeeded, in 1927 and 1928, in presenting a detailed assessment of the varied city-umland relationships and the principal processes of urban development.

Regional Transfer Systems

Bobek was one of the few urban geographers who actively challenged the commonly accepted, morphological definition of the city. He flatly declared this conception insufficient and misleading, in so far as it did not take into account other fundamentally urban characteristics, namely, a city's socioeconomic function. Although Bobek attempted to work out a compromise by integrating the "old" with the "new," his proposal clearly expressed the novel emphasis of urban geographic research:

> [The city is . . .] a larger settlement which constitutes the

> [operational] center for versatile economic, political, and cultural activities in a more or less defined area, and whose physical structure allows the recognition of a graded increase of characteristic elements from the fringes to the center. (1927, p. 216)

Bobek proceeded to identify typical urban activities which necessitate and essentially profit from such general allocation processes as agglomeration and centralization. The historical permanence of their manifestation and the overall degree of their functional specialization were considered to represent the cumulative consequences of urban growth and increased interaction between urban center and rural umland. Three elemental categories of urban activities were distinguished: (1) trade and commerce; (2) politico-administrative; and (3) cultural institutions and functions.

The city was to operate as the organizing focus for all three groups of urban activities, directing both the individual and the collective movements of goods, people, and information over an area the size of which depended on the type of commodity or service being offered by the city. The functional organization of both administrative and cultural institutions was described to be pyramidal in structure, resulting in a hierarchical sequence of dominance and subordination. The sector of trade and commerce was said to be organized in a similar manner. Yet, since it allowed greater flexibility of locational choice, it lacked the rigid organizational structure of the other two. As commercial center, the urban settlement was to monitor the fundamental interplay between producer and consumer, supply and demand. The major objective was to create and maintain the conditions of equilibrium between supply and demand, both on a local and on a regional level.

In evaluating the major processes contributing to the foundation and early development of urban settlements, Bobek acknowledged, as did Gradmann, the elemental importance of the evolving system of socioeconomic interaction between urban centers and rural surroundings. The mechanism of economic reciprocity, created by the principal division of labor between city and country, was noted as initiating early urban growth. Beyond the early developmental stages of local exchange markets, other factors had to be isolated to explain the increased diversification of functions exercised by the city and its rural umland. The formation of regional transportation and communication networks was viewed as providing the impetus for such a development. Bobek's analysis of the impact on the socioeconomic structure of urban center and rural area focused on the following general aspects:

> (1) The production of agricultural goods increased, as a greater number of producers considered it economically profitable to ship their produce to the nearest urban center. Economic distance became more decisive than mere geographic distance.
>
> (2) The functional organization of the urban center diversified. The range of urban influence increased, as did the size of the immediate rural hinterland.

(3) A well-developed, regional transfer system facilitated the balancing of eventually occurring interregional deficits in supply-demand conditions, distributing excess supply to deficit regions.

(4) With improved accessibility—that is, reduced economic distance—competition for trade and market areas between cities intensified. The relative location of a city within the transfer system determined whether it continued to grow or be dominated by the nearest larger city. Thus urban growth was defined as competitive potential generated by existing differences in transfer costs or, as Bobek put it, by variations in "traffic tension." In toto, Bobek asserted that improvements in the operational efficiency of the three basic transfer modes strengthened the functional and territorial coherence of city and umland and promoted additional urban growth at selected locations determined by the cumulative effects of transfer cost advantages.

Bobek's work appears to be the final proof of the growing preparedness of urban geographers to consider the various socioeconomic activities and their functional interdependencies. At first, Hettner's analogue model had advanced the concept of evolutionary struggle for economic viability, the pivotal factor being the maintenance of urban market and trade areas. Gradmann, then, had suggested that the historic course of politico-economic decisions exerted the prominent impulse for the differentiated rates of urban growth and the spatial distribution of urban settlements. Local exchange markets were described as the most stable elements in an expanding system of cities as induced by regulated, regional trade unions. Bobek, finally, asserted that the contemporary patterns of the spatial distribution and functional diversification of cities reflected major structural and operational variations in existing regional transfer systems. The city was defined as the organizational focus, so as to equalize regional differences in supply-demand conditions.

Although the "naive" description of observed geographic reality seemed to have prevailed in its most formal elements, it is evident that such exploratory investigations had one principal denominator that distinguished them from all other studies: the positive recognition of economic space or, less abstractly, the economic landscape. To substantiate this rather embryonic notion additional analytical research was needed. Urban geographers were to identify tractable variables indicative of major socioeconomic processes influencing the location, growth, and distribution of cities. Preceded by the formulation of a provisional working program, conceptual models were to be developed which were explanatory in nature and offered a plausible framework for predicting the probabilities of future events and developments. Once again, the geographer faced the dilemma of having to make the choice between simplicity and individuality, normative theory and descriptive model.

IV. RÉSUMÉ: CONTINUITY OR RENEWED CONFLICT

During the three decades that followed Schlüter's programmatic discussion, urban geography had not only experienced a rapid expansion, but acquired a respected position in human geography as well. It seemed as if urban geographers had finally managed to establish the necessary communality among themselves, thus terminating the occasionally rather frustrating search for substantive identity and relative conceptual stability. The principal cause was without doubt the eventual relaxation of the rigid conceptual and procedural organization that early geographers had demanded of geographic research in general. Free at last from the restrictive concepts and predetermined methods of proto-academic logic, human geography developed swiftly to become a well-recognized, innovative subdivision of the discipline.

Urban geography, on the other hand, had the unexpected opportunity of defining the very object of its research interests: the city. The concept of the cultural landscape as developed by human geographers was adopted as the basic conceptual principle. During this period two distinct schools of urban geographic research emerged:

(1) The *morphographic* method centered on the formal description of structural differences of the contemporary urban landscape, stressing individual, local accounts and discouraging regional or crosscultural comparisons. Socioeconomic characteristics were randomly included as explanatory variables (form-function-correlates).

(2) A second group of researchers promoted detailed historiographic investigations. This perspective, known as the *morphogenetic* approach, encompassed two interrelated sets of research interests:

(a) the internal structural differentiation of cities, as it reflected the sequence of historic events, be they sociocultural, political, or economic in nature;

(b) the course of historic changes, because it furnished some explanation as to the foundation, development, and spatio-temporal distribution of cities.

The concept of the urban landscape functioned as a stabilizing element. It provided the conceptual focus of consistent research in both morphographic and morphogenetic studies. The prominent emphasis on detailed observation and classification of minute structural differences and historical changes enlarged the urban geographer's reservoir of partial explanation and gave him a better understanding of local events and developments. The lack of a systematic framework of explanation, however, precluded a successful synthesis of the many diverse local findings for the purpose of establishing the organizing *Überbau* of retrospective explanation and prognostic evaluation. In other words, the procedures of active field investigation had been delineated, whereas the modeling structure of a general theory had yet to be defined.

The solutions, thus far proposed, had remained ambiguous and ineffective. Even Bobek's relatively explicit discussion of how specialized economic activities affect urban development had drawn little response. The prevailing concern of blending the conventional study of locality with the novel analysis of functional interdependency seemed, curiously enough, to have reinforced, rather than challenged, traditional proecedures (Dörries, 1930, p. 314). As the majority of urban geographers continued to ignore the embryonic notion of economic space, they could not possibly accept Christaller's line of thought, for he intentionally disregarded the established focus of urban geographic research and resolutely bypassed the logic of conventional explanation (Bobek, 1936). Exceeding Bobek's futile compromise, Christaller reduced geographical space, as described by his contemporaries, to an abstract, geometric plane, the elemental structure of which was derived from space-organizational principles of economic interaction (1933).

In view of the leading arguments, as applied in the central-place model, and the unyielding position taken by its adversaries, it seems worthwhile reiterating some of the major findings that had been suggested by studies exploring the then-considered marginal problems in urban geography:

(1) *Socioeconomic organization.* It was commonly maintained that the spatial and functional division of labor represented the basic condition for the emergence of urban settlements. With specialization increasing in each sector, three classes of city-located activities were discerned: trade and commerce, administrative, and sociocultural institutions.

(2) *Location, urban growth, and spatial distribution.* The location of cities was evaluated in terms of economic profitability of individual urban enterprises. It was hypothesized that similar economic activities seek locations of comparable space-organizational potential. Speculation upon the overall significance of external economies of agglomeration, especially urbanization economies, was eminent.

(3) *Transportation and communication networks.* The degree of accessibility to the next nearest urban place was determined by differentials in the operational quality of existing transfer systems. Geographic distance was replaced by economic distance.

(4) *Local city systems and regional systems of cities.* It was stipulated that urban settlements and rural surroundings form a functional as well as a territorial entity. The city is centrally located relative to rural production. The degree of interaction between urban center and rural umland and the eventual development of regional systems of cities was believed to be regulated by the combined effects of above listed factors.

In his review of preceding research activities, Christaller repeatedly stressed the innovative character of individual works. Yet he generally maintained that the greater number of studies had paid little attention to a set of problems which he judged to be of central importance to urban geography: the analysis of the principal factors that determine the location, number, size and growth, and spatial distribution of urban settlements. In his search for an operative system of analysis and explanation, Christaller turned to economic studies. He acknowledged the overriding significance of two theoretical constructs which were designed to detect and relate primal economic processes structuring optimal locational systems of agricultural and industrial production, those of von Thünen (1826) and Weber (1909). Since the space-organizational impetus of industrialization and novel agrarian policy was not entered into the proposed model, the general discussion centered solely on the sector of tertiary activities. Hence, in as much as central-place theory specified the spatial structure of areal demand and point supply of goods and services, it was to supplement and check the locational models developed by von Thünen and A. Weber (Christaller, 1938).

As if to underscore his intention of focusing on a specific category of enterprises, Christaller coined a series of new terms and, thus, avoided possible conflict with conventional geographic nomenclature. The interaction of basic socioeconomic variables—as manifested in the "primary system of central places"—was defined by three axiomatic principles which regulated the optional variations of functional and structural organization of central place systems under conditions of politico-economic constraint (figure 4).

(1) *Centralistic order.* It was postulated that to be the organizational center of a given region constituted the primal function of an urban settlement.

(2) *Functional exclusivity.* An urban center of higher order contained all tertiary functions and more, by comparison with the next lower level place. This discrete, classificatory ranking of central places might also be referred to as the principle of additive growth.

(3) *Abstract, geometric space.* The concept of abstract space was viewed a necessary condition so as to test and compare variations in spatial structure of regulated economic interaction. The complex organization of actual space was reduced to an isotropic, two-dimensional plane commanding a uniform density of nonurban population, an equal distribution of purchasing power, ubiquity of natural resources, and unrestrained accessibility.

Based on these axioms, Christaller was able to overcome numerous practical difficulties of hierarchically ranking urban centers, their functions, and complementary regions. The principles of centralistic order and functional exclusivity, in particular, supported the initial notion of a regularly progressing hierarchy of central places. Both, in effect, presupposed a rigid, hierarchical system of singular dominance and linear subordination, thus precluding development of crisscross patterns of interaction among its elements.

It thus is curious that central-place theory for all its inno-

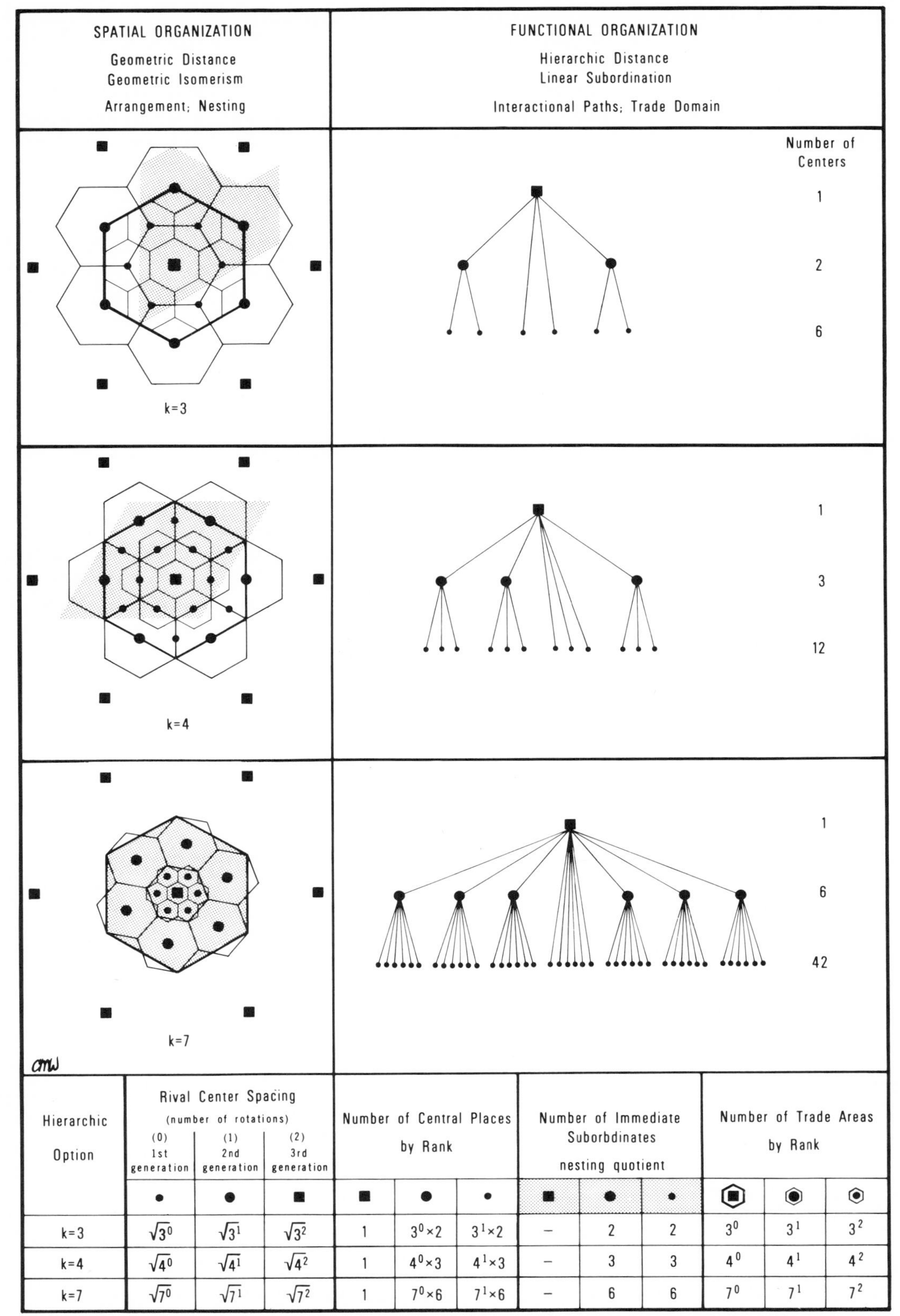

Hierarchic Option	Rival Center Spacing (number of rotations) (0) 1st generation	(1) 2nd generation	(2) 3rd generation	Number of Central Places by Rank			Number of Immediate Suborbdinates nesting quotient			Number of Trade Areas by Rank		
	●	●	■	■	●	•	■	●	•			
k=3	$\sqrt{3^0}$	$\sqrt{3^1}$	$\sqrt{3^2}$	1	$3^0\times2$	$3^1\times2$	–	2	2	3^0	3^1	3^2
k=4	$\sqrt{4^0}$	$\sqrt{4^1}$	$\sqrt{4^2}$	1	$4^0\times3$	$4^1\times3$	–	3	3	4^0	4^1	4^2
k=7	$\sqrt{7^0}$	$\sqrt{7^1}$	$\sqrt{7^2}$	1	$7^0\times6$	$7^1\times6$	–	6	6	7^0	7^1	7^2

Figure 4
The central place model: spatial and functional organization of a 3-tier, regular hierarchy – option k=3, k=4, and k=7 (after Christaller and Lösch)

vation is frequently understood as the final blow to the encrusted routines of traditional geographic research. Christaller, for one, was not inclined to contradict or negate the validity of both past and contemporary research programs. He wished rather to expand the conceptual range of urban geography by providing a model of systemized investigation and explanation which was to bring into focus the discipline's long-standing interests. Although "geographic reality" was cast into the rational dimensions of Euclidean geometry, central-place theory succeeded in introducing a novel perspective, which proved extremely effective in that it stimulated both empirical and theoretical research projects. By defining basic sets of economic interrelationships more precisely and relating them to space-organizational structures, Christaller's model contained standardized reference catalogs which facilitated procedures of field examination and provided incentives to further explanation and possible prediction. Most importantly, the study of central-place systems led to the formulation of four major areas within urban geography which were to attain focal character as research objectives: (1) the differentiation of discrete classes of urban settlements; (2) the investigation of prevalent modes of interaction between urban center and rural umland; (3) the evaluation of interurban and intraurban patterns of socioeconomic organization; and (4) the analysis of the spatio-temporal distribution of cities with regard to location, size, and function.

Thus, for the first time since Kohl's incisive deductions, urban geography was exposed to a systematic abstraction of the observed "real" world. Central-place theory set forth an integrative, conceptual framework which promised to open up new frontiers of research and laid the foundation for increased interdisciplinary activities. It signaled the end of a period of stagnation and sporadic exploration and set the stage for an era of continued progress and intermittent conflict in geographic knowledge as well as theory.

4

CHANGING THEORETICAL FOUNDATIONS OF THE GRAVITY CONCEPT OF HUMAN INTERACTION

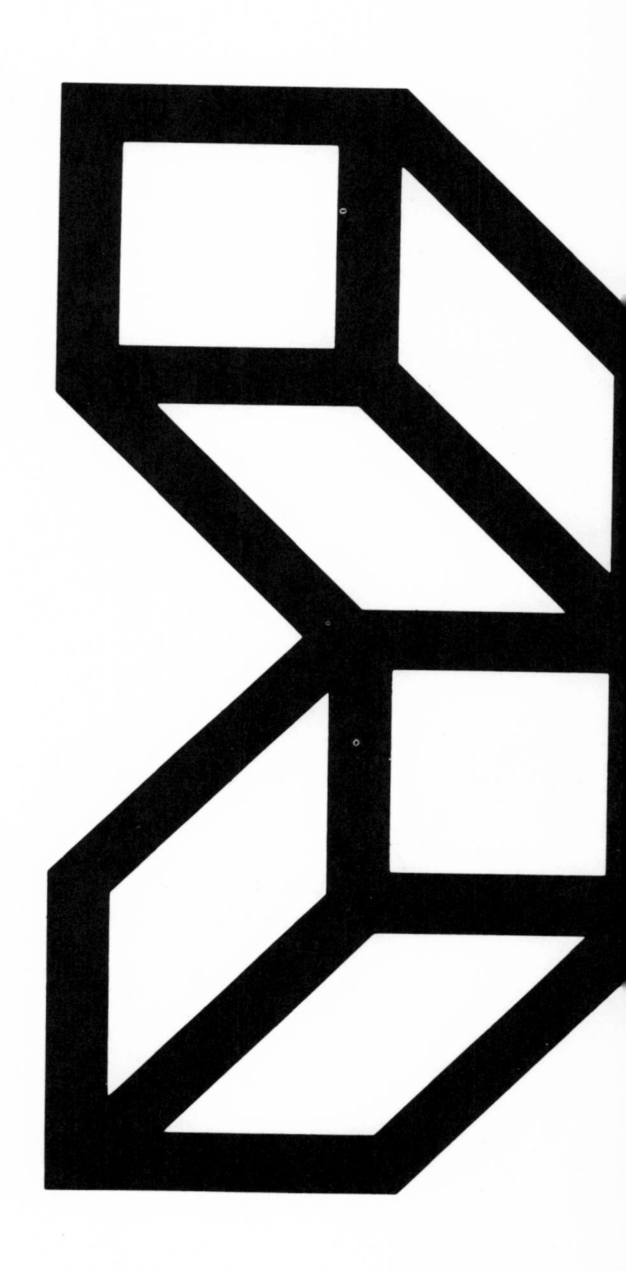

Thomas R. Tocalis
University of Chicago

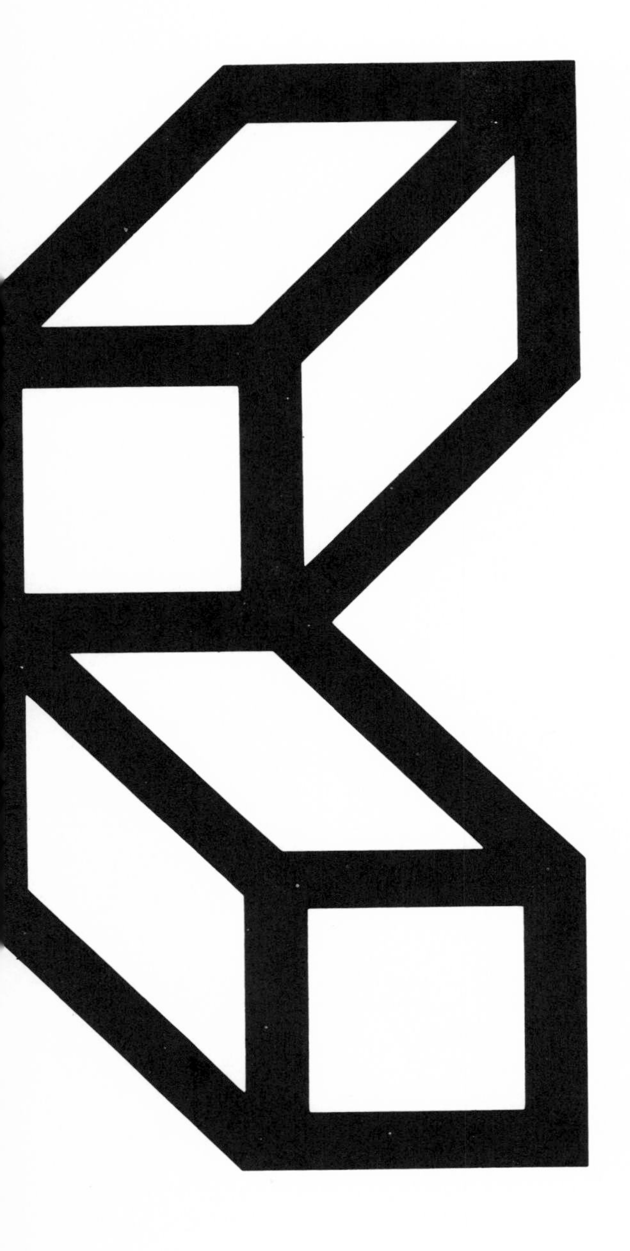

CHANGING THEORETICAL FOUNDATIONS OF THE GRAVITY CONCEPT OF HUMAN INTERACTION

Some geographical ideas have an astounding resiliency and are able to survive despite drastic changes within the discipline and in external stimuli. The gravity concept of human interaction is a case in point. Its progressive development can be said to parallel that of geography itself. Each phase of its evolution reflects the influence of the prevailing philosophies of science that are currently in vogue during each period.

This paper follows the course of the gravity model from its initial conception in the minds of Greek philosophers to its most current form. In this respect the paper is a review of the literature. It is hoped, however, that it is more—an attempt to interpret progress in geography through the analysis of one of the most extensively developed concepts within the science.

I. ORIGINS OF THE GRAVITY CONCEPT

An investigation of the early philosophical foundations of social science will provide insight into the emergence of the gravity concept of human interaction. The first formulations of group behavior were developed by members of a mechanistic school of social theory who interpreted social phenomena in the terminology and concepts of physics, chemistry, and mechanics. What resulted was social physics, the essential characteristic of which was a monistic conception of the universe, including the universal application of all natural law, or unity of all its law (Sorokin, 1928).

This belief in the universality of nature in both physical and social processes was first acknowledged by ancient Greek philosophers who were adherents of atomistic monism. The critical advance in mechanistic interpretations of social systems, however, was preceded by periods of scientific achievement in the physical and mathematical sciences. Thus the seventeenth century provided the necessary energy and impetus for the formalization of the mechanistic school of social theory. During this period much effort was expended in attempts to explain social phenomena in the same manner that mechanics had so successfully investigated physical phenomena. Sorokin summarized the essential characteristics of seventeenth-century social physics as follows:

> First, in contrast with the preceding thinkers, the social theorists of the seventeenth century (Hobbes, Spinoza, Descartes, Weigel, Leibnitz, and others) abandoned anthropomorphism, teleologism, moralism, hierarchism in their study

> of man's nature, mentality, behavior, and social phenomena. Second, they began to study social and physical phenomena as a physicist studies physical phenomena, rationally but objectively. Man was regarded as a physical object—a kind of machine or physical automaton. (1928, p. 5)

Society became an astronomic system whose elements were humans, bound together or separated by forces of mutual attraction or repulsion. Society was viewed as having evolved as a gradually enlarging series of force fields interacting with each other. Social theorists utilized the concepts of space, time, gravitation, inertia, and force to explain social activity. These social physicists viewed each human group as a state of equilibrium maintained through the opposition of centrifugal and centripetal forces.

II. EMERGENCE OF SOCIAL PHYSICS

While the efforts of seventeenth-century social theorists to form a science of social physics failed, their ideas were to influence later writers. Nevertheless, the mechanistic theories of the eighteenth and the beginning of the nineteenth century were merely restatements of the work of these earlier social theorists. It was not until the appearance of H. C. Carey's *Principles of Social Science* (1858) that social physics received a definite thrust forward.

In his monumental work, Carey explicitly formulated a gravity concept of human interaction, as follows:

> . . . It is scarcely possible to study these facts (physical phenomena) without arriving at a belief in the universality of laws governing matter, whatever form that matter may take, whether that of clay, coal, iron, wheat, or man—whether aggregated in the form of systems of mountains or in that of vast communities of men. . . . Here we have the great law of molecular gravitation as the *indispensable* condition of the existence of the being known as man. Man tends of necessity to gravitate toward his fellow man. The greater the number collected in a given space the greater is the attractive force there expected. Gravitation is here, as everywhere else in the material world, in the direct ratio of the mass, and in the inverse one of the distance.
>
> All the members of the human family do not tend to come together in a single spot of earth. Because of the existence of the same simple and beautiful and universal law by means of which is maintained the beautiful order of the system of which our planets form a part . . . each has within itself a local centre of attraction enabling it to preserve its form and substance, despite the superior attraction of the larger bodies by which it is everywhere surrounded. So it is throughout our world. Look where we may we see local centres of attraction towards which men gravitate, some exercising less influence, and others, more. The laws of physical science are equally those of social science, and with every effort to discover the former we are but paving the way for the discovery of the latter. (Carey, 1858)

These few paragraphs provided the underpinnings for the development of modern social physics. Nevertheless, despite this early formulation of a social theory based upon the universality of the natural laws of physics, empirical studies were almost nonexistent.

It was not until 1885 that E. G. Ravenstein presented empirical evidence which indicated that migration of workers tends to be toward larger cities and that as distance increases between the source of migration and the center, the volume of migration decreases. In 1924, E. C. Young proposed that the relative volume of migration between two centers varies directly with the "force of attraction" of the destination, and inversely with the square of the distance between the centers, a direct reference to the force of a magnetic field. Reilly (1929; 1931) presented the first explicit mathematical formulation of a gravity concept of human interaction in his study of the trade areas of 225 Texas cities. The study indicated that two cities draw trade from a smaller intermediate city approximately in direct proportion to the population (P_a, P_b) of the two attracting cities, and in inverse proportion to the square of the distance of each of the larger cities (D_a, D_b) from the smaller center. Mathematically stated:

(1)

$$\frac{B_a}{B_b} = \left(\frac{P_a}{P_b}\right)^N \left(\frac{D_b}{D_a}\right)^n ,$$

where (N) equals unity and (n) equals two. Reilly did not attempt to formulate any theoretical base for his model, and, as a result, it did not appreciably contribute to the development of a theory of demographic gravitation.

John Q. Stewart and the Codification of Social Physics

Extensive efforts at constructing a theory of human interaction based upon analogies with physics did not occur until the 1940s and 1950s when social phenomena and Newtonian physics were linked in the formulation of J. Q. Stewart. A professor of astronomy and physics, Stewart became the most vital advocate of social physics during this period. His entire career was marked by a continuous effort to supply a functional link between the physical and social sciences.

Transferability of Ideas

Stewart was constantly at odds with the position often taken that "scientists are prone to believe that every branch of study has its own peculiar relationships, most of which are not duplicated elsewhere" (Stewart, 1948a, p. 21). This idea of uniqueness had been previously opposed by Leibnitz (1890; 1896), who also firmly believed in the transferability of ideas and relationships from one science to another. The hesitancy by social scientists to utilize the laws of physics and their lack of attention to empirical regularities observed in social phenomena prompted Stewart to construct a conceptual framework for social physics.

Stewart argued that, in its most basic form, social physics seeks to discover mathematical order in human relations through the reduction of the raw data of social life into concise demographic laws. Social physics analyzed demographic, economic, political, and sociological situations in terms of the purely physical factors of time, distance, mass, and population. Stewart stated: "People can be counted. They do exist in space and time. Their activities are subject to mechanical limitations which can be described" (1948a, p. 20).

Unfortunately, the phenomena with which social physics dealt proved to be less tractable and more intricate than those of the physical world. "This is partly because we view them from the microscopic rather than the macroscopic level, being ourselves 'molecules'" (Stewart, 1952, p. 114). Stewart thought this proximity effectively prevented the social sciences from obtaining an observation point distant enough so that mankind appeared as a continuously flowing fluid upon the earth's surface. Not surprisingly then, this hindered the formation of social concepts at a macroscopic level.

Duality of Approach

Stewart's awareness of this myopic condition encouraged him to develop an interesting analogy which he believed would provide social scientists with a much-enlarged focal length for the investigation of man's spatial movements. For in effect

> . . . today's physics exhibits two realms of natural law—the microscopic in the newer quantum mechanics, the macroscopic in the old mechanics of Newton and Hamilton. May there not similarly exist for human relations the same two realms of natural law and justice—microscopic or personal, and macroscopic or social? (Stewart, 1952, p. 130)

Stewart chided the behaviorists and social moralists of the period, whose "continued insistence on examining people's purposes and motives only, blocks the way to a science of society as a whole, just as similar sentimentality and animism in Aristotelian physics made that physics useless century after century" (1948b, p. 32).

With this duality proposed, Stewart maintained that social physics of a macroscopic nature must be able to answer one crucial question: Can social physics sufficiently approximate an overall solution for mass sociological problems by an analysis that averages the myriad conflicting desires and the multitudinous characteristics of individuals into a uniform system based upon Newtonian laws of gravitation?

Principle of Indeterminancy

Stewart compared the activity of the complex, unpredictable, individual man with the activity of individual atoms and their uncertainties. This, of course, parallels the famous principle of indeterminacy stated by the German physicist Werner Heisenberg (1958; 1959; 1961). Heisenberg stated that the motion of individual molecules cannot be described with great precision. The averaged motions of individuals in a group conforms to mathematical laws, however. Stewart extended the Heisenberg principle to human motion.

With this parallelism established between man and molecules, Stewart (1948a) defined social physics as the approach that examines human relations in terms of space, time, and number. Its raw materials are statistical observations about people. Whenever appropriate, social physics draws on mathematical physics for suggestions and analogies which seem usable as guides for discovering and organizing social concepts. For Stewart, social physics contained six social dimensions: distance, time, social mass, social temperature, desire, and number of people. According to Stewart (1952), these six factors make the dimensional structure of social physics isomorphic with physics, thereby providing a reliable and complete analogy between any two or more situations involving like phenomena.

Analogies with Newtonian Physics

Stewart realized that the precarious position of social theories derived solely from carry-overs from the physical sciences. A detailed analogy is not sufficient proof. The resemblance must be more than verbal; it must provide a functional or abstract relationship that is common to both fields. The analogy must be supported by empirical tests using observed social data. The method of developing social theories deductively from considerations having a priori appeal must be replaced with a detailed process which searches for significant regularities among the mass of social data. This would provide for the vital need of social studies: better agreement, better coordination, a central nucleus of fact, and relations accepted by the great majority of experts (Stewart, 1950).

Stewart turned to Newton's laws of motions to provide the theoretical base for human interaction. Newton's original formulation stated that two particles of mass (M_i, m_j), one at point (i) and one at point (j) separated by a distance (d_{ij}) will be acted upon by a force of attraction (F) along the line joining them with a magnitude of

(2)

$$F = \frac{GM_i m_j}{(d_{ij})^2} ,$$

with the gravitational constant (G). The mutual energy (E) of the two masses within the gravitational field is

(3)

$$E = \frac{GM_i m_j}{d_{ij}} .$$

And the gravitational potential (V_i) which mass (M) produces at point (i) is

(4)

$$V_i = \frac{G_m}{d_{ij}} \quad .$$

Stewart substituted population for mass in Newton's first three laws. Thus the demographic force of interaction between group (N_i) at a distance (d_{ij}) from group (N_j) is

(5)

$$F_{ij} = \frac{G(\mu_i N_i)(\mu_j N_j)}{(d_{ij})^2} \quad ,$$

where (G) is constant of proportionality equivalent to the gravitational constant in physics. A molecular weight (μ) is attached to each population group (N). This is a constant whose value depends upon the group being studied.

Stewart was one of the first to show that the effects of distance on human interaction vary with the socioeconomic level of the population examined. Empirical evidence presented in Stewart's study of human interaction displayed a range in the value for the "molecular" weights from 0.8 to 2.0 in the United States. Stewart maintained that "although sufficient numerical data are not yet available, it is certain that the relative influence at a distance produced by a primitive savage is very much less than that of people in an advanced state of material civilization" (1952, p. 121).

Stewart theorized that the value of the "molecular" weight was a function of the population group's kinetic energy (K) defined as $K = Fh\mu$, where (F) is the mass of artificial goods moved per unit time, (h) is the average length of haul, and (μ) is the average speed from origin to destination. This demographic mass per capita measure of mobility was based upon the virial theorem of Clausius (1885; 1899) and was later applied by the French mathematician Poincare (1892; 1905; 1913) to a cluster of gravitational bodies. This theorem states that as the kinetic energy of a system increases, the tendency for the system to expand also increases. This theorem can measure the effect that the technological state of communications has upon human interaction, "for the relations of people to one another require more than people to support them; they require also national resources and technological facilities" (Stewart, 1942, p. 57).

Stewart's second major formulation defined "demographic energy" as the energy produced by the mutual gravitation of two populations (N_i, N_j) spaced at a distance (d_{ij}), where the amount of human interaction between these two centers is:

(6)

$$E_{ij} = \frac{G(\mu_i N_i)(\mu_j N_j)}{d_{ij}} \quad .$$

Thus, when one aggregation of people is within another's gravitational field, the result is a strong interaction between the two, and if the populations are similar in composition, there is a mutual tendency to coalesce. This demographic energy is directly related to the product of the masses and inversely to the distance separating them.

Stewart's last transformation of Newtonian law was that of the "potential of population." This is a measure of the intensity of the possibility for interaction. The potential influence on an individual (iV_j) located at (i) by a force generated by a population at (j) is:

(7)

$$iV_j = \frac{GN_j}{d_{ij}} \ .$$

Stewart admitted, however, that his formula is somewhat modified by an extra energy of "cohesion" found within population centers and is in part a function of population density.

Stewart introduced two other concepts derived from physics: social temperature and social entropy. Social temperature measures the rate of activity of a social mass relative to its interactions. Social entropy is an even more abstract concept. For Stewart, it "refers to the number of different arrangements of the social molecules that leave the assemblage or community unaltered with respect to the macroscopic variables, which may be taken as the social temperature and the population density" (1952, p. 28). Analogous to physics, social entropy increases with the level of activity, but tends to decrease with increases in population density. Due to the difficulty of defining the complex variables, these two abstract concepts developed by Stewart were not given a rigorous mathematical interpretation until the work of A. Wilson in the late 1960s.

Nevertheless, Stewart's analogies formed the basis for much of the following empirical investigation of human interaction. His concept of demographic gravitation was a highly sociological generalization. Although Stewart emphasized analogy with mechanical physics as an aid to research, he did not diminish the conventional and necessary emphasis on seeking agreement between hypothesis and observation. He foresaw the need of social scientists to search through vast arrays of sociological data, seeking empirical regularities expressible in the mathematical terms of Newtonian physics. And projecting the future effect of distance upon human interactance, Stewart envisioned that "when social physics advances to the quantum level, it may appear that the separating effect of distance is only external and statistical, and that one person can come in contact with another through impulses which make nothing of space and perhaps nothing of time also" (1948a, p. 23).

Stuart Carter Dodd and the Development of Sociometry

Contemporaneous with Stewart were several other writers who were involved with formulating gravity concepts of human interaction based on somewhat differing assumptions; among them was Stuart Carter Dodd.

Stuart Dodd's major concern was with developing mathematical formulations for "social forces" in order to construct a theoretical foundation for "sociometry." For Dodd (1943c), sociometry was that subfield of sociology which examined human interactional phenomena through the analysis of people responding to people. These responses were expressed as pairwise interactions within a matrix format. His search for basic relationships in human interactions was influenced by the writing of another sociologist, J. L. Moreno, who observed that

> . . . mankind is a social and organic unity—tendencies must emerge between the different parts of this unity drawing them at one time apart and drawing them another time together. . . . These attractions and repulsions or their derivatives may have a near or distant effect not only upon the immediate participants in the relation but also upon all other parts of the unit which we call mankind. . . . Its organization develops and distributes itself in space apparently according to a law of social gravity which seems to be valid for every kind of grouping irrespective of the membership. (Moreno, 1947, p. 288)

Thus Moreno introduced the major components of human interaction—population, time attraction, repulsion, and distance. Dodd modified this somewhat when he dealt only with a "net social force,"—that is, "that part of the stimulation which produces an acceleration of a social change in a population in addition to overcoming whatever resisting forces there may have been in the total situation" (1936b, p. 61).

The Interactance Hypothesis

In many respects, Dodd's approach in the development of his "interactance hypothesis" is similar to that of Stewart. Dodd relied on a direct analogy with mechanics to develop the basic concepts of human interaction. His primary concern was the functional similarity between Newton's laws of gravitation and human interaction. Dodd was quick to point out, however, that the functional similarity does not assert any metaphysical assumptions about the nature of social phenomena but was utilized as a frame of reference. Within this framework, Dodd was able to define analogues of demographic gravitation as compounds of three basic factors: time, population, and indices. An index measures any quantitatively expressible characteristic of a population.

In order to define analogues of group gravity, Dodd (1951b) assumed that human beings, although highly variable and complex when observed in mass, conformed to certain regularities very similar to the regularities of molecular activity observed by physicists, who have always studied molecules in large masses, never as identifiable individuals. A dependence upon the averaged actions of man displaying regularity requires gathering large samples in order to eliminate random fluctuations or other factors from overriding gravity principles. The relationship between sample size and the validity of any study of human interactance prompted Dodd to state that "sociological laws should most often be stated in terms of sociological probability. That sociological laws, or for that matter all empirically derived laws in any science, are matters of probability, to be expressed with probability or correlation coefficients ranging from zero to unity, seems a basic postulate" (1943a, p. 179). Dodd's application of probability theory in defining an interaction matrix, combined with his utilization of dimensional analysis in the formulation of the interactance hypothesis, proved to be his greatest contribution to the gravity concept of human interaction.

The hypothesis of group gravity, or the "interactance hypothesis," seeks to predict the number of interactions of any one specific kind among people observed in groups. The elementary dimensions of time, space, population, and level of activity are the factors of interaction and of all human behavior which can be combined into many compound concepts. Dodd (1950) expands the terms of physics: time (T), mass (M), and distance (L) into the dimensional sociological factors of time (T), population (P), distance (L), and a characteristic of the population (I). The interactance hypothesis is then derived from the following compounds: Social Force expressed as $PI/T^2 = F$; Social Energy in the form $PI^2/T^2 = E$; and the Energy of Interaction of Human Masses, or $P^2I^2/T^2 = KP^2I^2/L$, the interactance hypothesis through dimensional analysis, relies on the single premise that phenomena can be described by a dimensionally correct equation among certain variables (Langhaar, 1951).

Role of Dimensional Analysis

Perhaps Gukhaman (1965) gives the major reason why Dodd may have utilized dimensional analysis in his investigation of demographic gravitation. When the number of variables is large, it is very difficult in practice, or even quite impossible, to reduce the results of the solution to a systematic form, to find the hidden relationships, and to combine these relationships into general quantitative laws. Thus, in order to supply sociologists with a theory of social gravitation in operational and verifiable terms, Dodd expressed interactance as the product of measurable power factors with their dimensional implications. In dimensional analysis, one physical quantity can only be related to another relevant physical quantity in some definite class of phenomena, by simple power laws, when there is a measure of similarity in the phenomena for all the systems considered (Duncan, 1953). This dimensional analysis forms the operational basis of quantitative empirical science.

Dodd first measured those variables he suspected of being important in the explanation of demographic gravitation. These variables were then placed in a data array. The relationships between these quantities, expressible in mathematical terms, then were sought. The analysis focused on the independent variable in terms of the relevant dependent variables as in the general form $Y = f(X_1, X_2, X_3, \ldots, X_n)$, or for Dodd's interactance hypothesis, the expected interactance $I_e = f(P, L, T, I)$ and $k = f(L, T)$. The more complex dimensional form which Dodd derived for his interactance hypothesis became:

(8)

$$I_e = kI_A^i \cdot P_A^p \cdot I_B^i \cdot P_B^p \cdot T^t \cdot L^{-\ell} \; .$$

The properties of the interactance constant (k) were isolated by using dimensional analysis, and Dodd showed that (k) had the dimensions of acceleration, a rate of change of a rate of change, and was equal to (LT^{-2}). This gravity constant changes with each type of interaction for one population at one time. Dodd was not able to determine how much this "constant" will vary with changes through time.

The constants (I_A) and (I_B) represent the per capita level of activity. They are, in effect, weighting factors used to balance the heterogeneity of the groups interacting, similar to the "molecular weights" of Stewart. These indices are constant for each group and are functions of various socioeconomic factors such as sex, age, income, education, and others. Dodd explicitly indicated that socioeconomic factors influence the degree of human interaction between nonhomogeneous populations.

The factors (P_A) and (P_B) represent the populations of two interacting groups. Dodd defined the relationship between interacting groups through an interaction matrix. The rows and columns of the matrix represent frequencies of interaction within the group. The law of joint probability states that the relative cell frequency is proportional to the product of its rows and columns relative to interaction frequencies. Thus the probability of interactance between (P_A) and (P_B) is:

(9)

$$\frac{P_A}{P} \cdot \frac{P_B}{P} = \frac{P_A P_B}{P^2} \; .$$

Thus, each pairwise interaction has a probability proportional to ($P_A P_B$).

Radiant Zone Concept

Dodd (1950) explained the inverse distance relationship in terms of a radiant zone concept which he believed applied similarly for Newton's law of gravity, demographic gravitation, and social diffusion from a point. He assumed that a person has a constant amount of energy produced by (social influences) which is equally likely to diffuse at an even rate in any direction. The areas of concentric rings vary directly with its distance (radius) from the emitting particle. Since constant energy is assumed to diffuse evenly, energy per unit area will decrease as distance from the emitting particle increases and will vary at a rate of (L^{-1}). Thus, the proportion of people who will be influenced in each zone is inversely related to their distance from the origin. Dodd was aware of the fact that if population density is not uniform, some function of the distance other than the first power (L^{1}) may give a better fit. In fact, he recognized the value of replacing absolute distance with some function of distance, such as travel time or travel cost.

Dodd's "interactance hypothesis" was not new. Nevertheless, its exact operational definition in observable indices with computable constants based upon its dimensional formulation was original. It explained human interaction as the product of seven measurable factors. Its probabilistic nature was acknowledged and was expressed in terms of correlation coefficients. His initial attempts at deriving interaction relationships through the laws of joint probability will appear again in the works of later investigators, as will his formulations based upon the product of power factors. For Dodd, the interactance hypothesis was but part of a more generalized law of interaction in which independent particles interact with each other, and any pair will interact in proportion to the product of the number of particles in each aggregate, and inversely to the distance between them. "If the particles are molecules, that principle becomes the law of physical gravity; if the particles are persons, the principle states the mathematical law of joint probability" (Dodd, 1951b, p. 234).

George Zipf and the "Hypothesis of the Minimum Equation"

It was also during this formative period that George Kingsley Zipf developed his "hypothesis of the minimum equation," a bold attempt at providing a unifying social principle of human interactance. The manner in which Zipf devised his concept differed from that of Stewart and Dodd in both scope and basic assumptions, although within his framework, Zipf was able to provide valuable insights to the concept of group gravity.

Unity of Natural Law

Zipf was a firm believer that social scientists too often ignore the inherent interrelatedness of social and natural sciences. Zipf assumed that the unity of natural law governs the behavior of the solar system, all living organisms, and the emotional and intellectual activities of the mind. All matter moves through space and time according to the natural laws of the universe. Thus, social scientists, in effect, examine certain details of universal activity. All principles that have been disclosed as applicable to the rest of the universe will apply automatically to social phenomena. We know that social phenomena are subject to the strictures of electro-chemical law and gravitation, because social phenomena involve matter (Zipf, 1942). This assumption allowed Zipf to construct functional links between social phenomena and Newtonian physics, and to investigate the organization of matter (in particular social organization) as a single problem in dynamics.

Principle of Least Effort

Of critical importance to Zipf's theory was his observation that in describing any phenomenon, two superlatives make the solution of a problem indeterminate and meaningless. Thus, a problem in dynamics can have only one maximum or minimum, and not more than one (Zipf, 1942). Nevertheless, should a particular superlative be found to apply anywhere in the universe, it would then by extension be inherent in all other natural phenomena. Since natural phenomena are known to move along paths of least effort, so must social phenomena. If the movement of matter in the space-time continuum is governed by the minimum least-action, then neither time nor space is being minimized as a requirement for movement of matter. For Zipf, a straight line path is only a special case of what we may call a least-action distance, or a minimal energy distance (1942). Thus, Zipf, by assuming the unity of nature and extending the principle of least-action, was able to anticipate the existence of a social theory that explains the size and number of communities spatially distributed so as to minimize total human effort—"the minimum equation."

Although Zipf's theory of social interaction is based primarily on the premise that a biosocial system seeks to minimize total work, it also is reducible to Newton's laws of motion. A social system contains a population (C) which is distributed over a large area. Zipf assumed that there is equality among individuals in production and consumption of goods and services. Therefore, the biosocial system's spatial structure is a function of three variables: the terrain, the diversity of raw materials needed, and the processes utilized by the system. As the system's need for a variety of raw materials (m) increases, the probability of any one location containing these materials (M) decreases rapidly; that is,

(10)

$$P(M_1)(M_2)(M_3) = P(M_1) \cdot P(M_2) \cdot P(M_3) \ ,$$

with the assumption that raw materials are randomly distributed. Zipf refers to this phenomena as the "area-diversity" relationship—the greater the area, the greater the probability it will contain a larger variety of natural resources.

Zipf noted three effects of an increase in demand for a greater variety of raw materials upon the social system. As the number of raw materials (m) increases, the rate of movement of (m) also increases. There is an increase in the diversity of processes needed to produce a variety of goods and services as well as the development of specialized sectors of the economy. This expansion in the complexity of production and consumption directly influences the number (n), the population size (P), and the spacing (D) of cities. When considering a basic economy using only a few localized raw materials, the most efficient operation is obtained at the sites of the raw materials. Therefore, the system is composed of a series of small communities equally spaced in relation to raw materials sources.

As the complexity of the social system expands, however, points of production are no longer points of consumption. Exchange of goods and services takes place between centers of production. Zipf thus noted that "all persons and all communities will be located with reference to this two-directional flow of traffic, which, in turn is subject to the exigencies of the minimum equation" (1947a, p. 628). Thus, the sum of the products of all masses moved into production, through production, and back to consumer, multiplied by their respective work-distances, is minimized. Work-distance in this context refers to the amount of human effort necessary, directly or indirectly, to move the mass of products over the respective distance.

It is at this point that conflict between man's dual role of producer and consumer arises. Producers minimize total effort by locating at the source of the raw materials; they are governed by the "force of diversification." On the other hand, consumers benefit through the concentration of goods and services, as they submit to the "force of unification." Within a closed system, the force of diversification creates a larger number (n) of communities with small populations, while the force of unification creates a smaller number of communities (m) with larger populations. Zipf knew that when these forces were observable on graph paper with double logarithmic coordinates, the relationship appeared as a straight line. The relationship is stated as follows: Any number of cities (m) ranked (r) according to decreasing size (P) will follow the equation $r \cdot P^q = K$, where (r) is a positive integral and the exponent (q) is the ratio of the magnitude of the force of diversification to the force of unification. This relationship was also expressed by Zipf as a harmonic series in the form:

(11)

$$P \cdot Sn = \frac{P}{1^k} + \frac{P}{2^k} + \frac{P}{3^k} + \cdots \frac{P}{n^k},$$

where (P · Sm) equals the total populations of the system; (P) equals the population of the largest city, and (p) equals (1/q). Zipf concluded that the harmonic series is closely related to the ratio of the inverse square, which in turn is inherent in three-dimensional space. We know that the inverse square relationship is venerable in physics (for example, gravitation, electricity) and is found notably where action is involved. We should not be surprised to find evidence of it in biological and mental phenomena where action is also involved (Zipf, 1942).

P_1P_2/D Hypothesis

With this relationship established, Zipf investigated the nature of interaction between pairs of communities. All production and consumption are assumed to be internal, with equal income and employment throughout the closed system. Let (C) represent the total amount of goods (over a given time interval) distributed evenly within the system over a given time interval. Then a city with population (P_1) will receive and send goods in the proportion (P_1C); a city of size (P_2), in the proportion (P_2/C). The amount of interaction between (P_1) and (P_2) is then ($P_1 \cdot P_2/C_2$), ignoring the factor of least distance (D). Zipf introduced the inverse relationship of least distance, since the probability that any kind of goods in (P) is directly proportional to (P); the probability of finding it there from a distance (D) is inversely proportional to (D) by precise geometric probability (in the sense that the angle subtended at a point by a given disc is essentially inversely proportional to (D) distance from that "point"). The situation is not altered by substituting work distance (D) (Zipf, 1947a). Therefore, the interchange of goods between any two communities, (P_1) and (P_2), will be

(12)

$$Y/C - \frac{P_1 \cdot P_2}{D} .$$

These are the basic formulas of Stewart's demographic gravitation and Dodd's interactance hypothesis, if the population is assumed to be homogeneous. The single difference is that for Zipf, (D) is the distance that minimizes the total work flow between centers (P_1) and (P_2).

The (P_1P_2/D) hypothesis is but a corollary of Zipf's more general theory concerning the spatial distribution of communities based upon the minimization of total work. Zipf clearly indicated that there are two forces at work within his biosocial system: one of attraction and one of repulsion. Zipf also acknowledged that the present ($P_1 \cdot P_2/D$) relationship (a two dimensional "gravitation"), like the three dimensional gravitation of physics, though theoretically rectilinear and doubly logarithmic, can be modified in respect to slope and rectilinearity by other factors (1947c). The exact nature of these factors is left for subsequent investigators.

Stouffer's Intervening Opportunity Theory

These concepts of human interaction were based on the complex forces related to cost of movement and number of persons available to move—the "size-distance" or "gravitational theory." A second theory, however, states that movement results from socioeconomic imbalances between communities, often referred to as the "socioeconomic push-pull theory" (Isaac, 1947; Goodrick, 1936; Thomas, 1942). An interesting attempt at combining the two theories was undertaken by S. Stouffer in his development of the "intervening opportunity" model.

Mobility and Distance Relationship

The most unusual aspect of this theory is Stouffer's assumption that mobility was not necessarily a function of distance. It proposes that the number of persons going a given distance is directly proportional to the number of intervening opportunities (Stouffer, 1940). The theory takes the form

(13)

$$\frac{\Delta Y}{\Delta S} = \frac{a}{x}\frac{\Delta x}{\Delta S} \ ,$$

where (ΔY) equals the number of persons moving from an original point to a circular band of inner width ($S-1/2\Delta S$), and an outer width of ($S+1/2\Delta S$), with Δx equal to the cumulative number of intervening opportunities between the origin and the distance (S), and Δx equal to the number of opportunities within the band (ΔS). The relationship between distance and mobility is an indirect one; assuming the random distribution of opportunities, the cumulative number of opportunities is a function of distance.

Stouffer continued his analysis by assuming a continuous function for the number of intervening opportunities. Thus (Δx) becomes (x - $f(S)$), a function of distance. Substituting this value in the original equation produces

(14)

$$\frac{dy}{dx} = \frac{a}{x}\frac{dx}{dS}$$

and integrating results in $Y=a \log f(s)-C$. Thus the number of migrants (y) who stop at any point within a circle of radius (s) is directly proportional to the logarithm of the number of opportunities within the circle. Stouffer realized, however, that a migrant's detailed knowledge of the number of opportunities decreased with an increase in distance from the opportunities. In order to account for this effect of distance upon perception of opportunities, Stouffer defined apparent opportunities (z) as a function of actual opportunities (x), postulating that as (x_i) increases, (z_i) increases at a slower rate. Therefore, the number of persons who will move somewhere within distance (S_i) is directly proportional to

(15)

$$z_i(Y_i - K^{z_i}) \ ,$$

where $z_i=m \log x_i-C$. Unfortunately, Stouffer was so overwhelmed by the difficulty of determining the exact relationship between distance and the perceived opportunities that he reverted to the original formulation, using the actual number of opportunities.

Stouffer also postulated that in a spatial system where all opportunities are distributed continuously in space in the form $x = ks^b$, the previous log formula would be altered to take the form $y = a \log s + C$, a logarithmic straight line. Nevertheless, Stouffer's data on the movement of households in Cleveland displayed a distinct curvilinear relationship, becoming asymptotic as distance increased. This supported the original theory of intervening opportunities:

(16)

$$\frac{\Delta y}{\Delta S} = \frac{a}{x}\frac{\Delta x}{\Delta S} .$$

Empirical Results of Bright and Thomas, Isbell, and Strodbeck

A series of studies utilizing Stouffer's intervening opportunities theory quickly followed. Bright and Thomas (1941) discovered that an excess of expected net migrants occurred in the first two distance bands. The authors attributed this to the fact that while the model did not directly postulate a relationship between mobility and distance, the model overestimated, because of the absence of distance in the first band, opportunities without "intervening opportunities." A study conducted by Isbell (1944) found that major discrepancies occurred and observed frequencies exceeded the expected in distance intervals containing large cities. Isbell concluded that opportunities may not be in direct proportion to size, and individual opportunities may have a different level of attractiveness—a possibility ignored by Stouffer. Bright and Thomas also observed that, since Stouffer measured opportunities in terms of distance bands, his formulation could not account for directional flows of migration, as all opportunities were obviously not "intervening opportunities." Isbell's study of migration in Sweden indicated that socioeconomic variables also influenced individual mobility—a factor not accounted for in Stouffer's formulation.

Nevertheless, the major problem was pointed out by Strodbeck (1949), who stated that a criterion for a definition of opportunity is required. This criterion is difficult to develop and, in general, tends to apply to individual case studies only. In spite of these major shortcomings, however, Stouffer's theory of intervening opportunity introduced a "social" distance to the gravity concept of human interaction. This theory pointed out that for every human movement, some type of goal or satisfaction is sought, and the probability of this desire being fulfilled increases with intervening opportunity and distance. The theory's close relationship with the pure gravity model insured its continued appearances in human interaction studies as an integral part of gravity models.

Summary of the Initial Phase of Development

With the completion of the pioneering works by Carey, Stewart, Zipf, Dodd, and Stouffer, the first phase of the development of the gravity concept in human interaction was concluded. This unique period was characterized by the following:

1. The attempt was made to explain human interaction in terms of universalities both in theory and in application.
2. The laws of Newtonian physics were applied directly or indirectly to the explanation of human interaction.
3. Interactions of men were measured in aggregate terms.
4. Individual behavior was not a subject of investigation.
5. Empirical studies were limited in number and scope.

Despite the paucity of empirical studies, this first era maintained a unity of approach not found in subsequent phases of the evolution of the gravity concept of human interaction.

III. THE PERIOD OF EMPIRICAL STUDIES AND THE SEARCH FOR APPLICATIONS

In contrast to the emergence era, the following period in the development of the gravity concept fell into two phases. The first briefer phase consisted almost wholly of detailed empirical studies testing the general hypotheses developed in the preceding period. The major portion of this second phase, however, witnessed myriad attempts at forming "composite" gravity concepts of human interactance, integrating various related concepts with that of the gravity concept. Probability theory, economic theory, and behavioral theory were all given spatial interpretations within the classical concepts of Newtonian physics. Without exception, these investigations did not seek universal concepts, or "laws" of human interaction. Instead, theorists attempted to formulate predictive urban transportation models based upon the analysis of interregional flows.

The Role of Metropolitan Transportation Studies

One of the most important factors influencing the development of the gravity concept at this time was the rapid proliferation of major urban transportation studies in the late 1950s and the early 1960s. Metropolitan regions, faced with expanding suburban periphery coupled with increasing rates of vehicular traffic, were in dire need of regional highway planning. Traffic engineers sought operational trip-distribution models, in an effort to analyze and predict future volumes and directions of interzonal traffic flows.

The most influential of these studies, in terms of the gravity concept, were the Detroit Metropolitan Area Traffic Study (Carroll), the Chicago Area Transportation Study (Carroll, Schneider), the Baltimore-Washington Interregional Transportation Study (Voorhees, Ikle), and the Penn-Jersey Interregional Transportation Study (Harris, Tomazinis). A critical feature of each of these studies was the manner in which human interaction was viewed. This was clearly stated by the Chicago Area Transportation Study:

> It is a basic theory of the study that there is order in the travel behavior in urban areas which can be measured and described. This order provides the basis for intelligent forecasting which is necessary so that solutions will cope with problems of the future—not just the present or the past. (1959, p. 15)

This search for order led to the development of three basic trip-distribution models. The first, the gravity model, was used to represent pairs of zonal interchanges independent of each other. The model was calibrated so that travel from any one node was affected by service and the attractions of all other nodes. The second, the intervening opportunity model, minimized travel from point to point under the assumption that each destination considered had a stated probability of being acceptable. Basically, the model stated that a trip is made to the closest acceptable location, regardless of time, cost, or distance. The third, a linear programming model, considered that zonal interchanges minimize costs, subject to the constraints of the system capacity (Meyer, 1971).

Nevertheless, despite the variety of methods utilized to forecast zonal interchanges, one concept remained common to all the traffic studies of this period. This characteristic was pointed out later by Chapin when he stated:

> A conversion from the micro level view of activity systems to a macro level perspective is essential if the foregoing schema (a conceptual model of travel behavior) is to be operationalized, and a macro approach calls for certain compromises if the schema is to be fitted to the real world. First of all, the changeover to a macro type involves an aggregative approach. . . . (1968, pp. 83-84)

A macro approach was not without its difficulties, as the subsequent investigations of traffic specialists were to indicate. Nevertheless, the goal, an operational model of human spatial interaction, was sought.

Traffic Studies of Ikle

The short transitional period consisted mainly of empirical studies concerned with examining two aspects of the gravity concept: the measurement of the "mass" factor-population, and the value and function of the distance exponent. Ikle (1954) objected to the attempts of past sociologists in seeking "universal" theories of human spatial behavior based on limited data—particularly Zipf's assumption of a linear relationship between frequency of interaction and distance. Ikle's objection was that the social and economic compositions of city populations differ substantially. Such composition differences affect the number of interactions between city pairs, because the frequency of interaction between two persons over a given distance depends on their social and economic characteristics (Ikle and Hammer, 1957).

As a result of his objections, Ikle introduced his version of "molecular" weights assigned to each city examined, much in the same manner as those of Dodd. The propensity to interact is reformulated.

(17)

$$H_{ij} = P_1 P_2 / D_{ij}^{b}$$

becomes $H_{ij} = K(P_1W_1)(P_2W_2)(D_{ij})^{-b}$ where (b) is the distance exponent, (D_{ij}) is the distance between the centers (P_1) and (P_2), and (W_1) and (W_2) are the socioeconomic weights attached to their respective centers. These weights also reflect the geographical location with respect to transportation and communication networks—a measure of accessibility. Ikle and Hammer assumed that "the frequency of interaction between two persons can be expressed as a power function of distance, with a negative exponent (-b), where (b) assumes various values. Data fails to justify the inverse linear or inverse square 'law' which previous investigators had suggested for the distance function" (1957, p. 314).

This is shown in Ikle's 1954 traffic study, where he utilized the logarithmic function $\log H_{ij}/(P_1P_2) - \log K^{-1} \log(D_{ij})$ in order to use the least squares technique. Iteratively deriving the values of (b), he found that the value of the distance exponent varied with the mode of travel. For air travel (b) $= -1.07$, but for auto travel, (b) ranged from $-.689$ to -2.57. The value of (b) also varied according to population density of the area under study. For interurban travel by car, (b) $= -2.57$, while intraurban travel by car resulted in a value of (b) $= -.689$. In this study Ikle was unable to discover a consistent pattern for variations in the distance exponent. Thus, he concluded that

> . . . the mechanism by which spatial distance is related to the frequency of human interactions is very involved and cannot be explained by some simple sociologic or economic interpretation. The cost of interaction over distance is obviously a factor which reduces the frequencies as distance increases. Furthermore, distance affects the probability that a relationship between two persons will become established. (Ikle and Hammer, 1957, p. 315)

This implicit referral of the role of distance in the probability of human interaction encouraged further investigations by Isard (1956); Schneider (1959); Harris (1954); and Huff (1963).

J. D. Carroll's Study

Although J. D. Carroll's 1955 study of telephone exchanges between Michigan cities lacked Ikle's probabilistic insight, it produced similar conclusions. The influence of an urban center upon its hinterland was determined to be a function of the city's size and distance from surrounding zones. Carroll assumed that the number and diversity of goods and services offered by any central place to be a function of that city's size. Thus, by inference, "the greater the variety and number of functions any city performs, the greater the probability that similar functions will not be provided in the adjacent area and therefore persons and organizations in the hinterlands will make use of such special functions" (Carroll, 1955, pp. b-4). This is a valid assumption, however, only if minimum threshold requirements exist in the hinterlands.

Using telephone data to measure urban influence, it was found that the distance exponent again varied—in this case from a value of −2.23 to −3.33. Although not noted by Carroll, there appeared to be a tendency for the distance exponent to decrease in absolute value as the city size or function—for example, state capital—increased. This indicates that the larger the city, the less its interactions are influenced by distance—an implication that the "mass" factor of population may require an exponential form (P), as suggested by Dodd. The difficulty of evaluating variable exponents for both distance and city size raises the question: What factors explain the variable exponents? Are they a result of absolute city size—the relative size of cities in pairwise interactions, or characteristics of intervening distances? Carroll does not provide an answer, for he insisted on stating that there appears to be a relatively uniform effect of distance on trip volume in all cases, and "influence tends to diminish according to a figure approaching the cube of distance" (1955, pp. D-10). This is an example of the same spurious logic that led Reilly to assume an inverse square relationship between distance and interaction.

Work of Voorhees

Voorhees (1957) acknowledged that many adjustments are required in the basic gravity model in order to analyze human interaction in urban areas. He said "for each trip purpose different indices had to be used to express the mass; and the distance had to be raised to some particular power which was in line with the trip under consideration" (Voorhees, 1957, p. 131). Distance raised to the second power was satisfactory for low- and middle-income groups in determining their shopping habits. This same exponential value did not apply to higher income groups. The variability of mobility among social groups must be incorporated within the function of distance.

In spite of varied results, this melange of empirical studies exhibits one unifying characteristic—each applies the classical form of the gravity model. This is not the case with subsequent investigators. The validity of applying the classic gravity concept to human interaction comes under a barrage of criticism on both methodological and theoretical grounds.

Objections of Carrothers

A major critic, Carrothers (1956a), objected to the almost dogmatic adherence to the physical laws of matter in Newtonian physics, an adherence he felt hindered the development of more viable concepts of spatial interaction. Carrothers acknowledged that human beings in large masses interact in ways subject to mathematical analysis and definition. But he seriously questioned the theory that this interaction follows Newtonian laws of physics. In fact, he stated, "I would indeed go further and take the position that the fundamental differences between man and his component molecules completely invalidates any arbitrary insistence upon the application of the laws of physics, as such, to the analysis of human interaction" (1956a, p. 10).

Carrothers also detected a hesitancy on the part of some authors to consider more than the variables of population and distance in their studies. The difficulty, he postulates, is one of finding explanatory variables which are both significant and measurable. The time variable presents such a problem. Carrothers admitted that while time and distance are obviously related, there are occasions where time may be the significant factor of friction in human mobility. In these cases, Carrothers believed it advantageous to treat time separately as an explicit function.

Even the two variables of distance and population were viewed as being more complex than had generally been thought. Carrothers was not convinced that distance was a simple exponential function. In fact, evidence put forth by Dunn (1956) indicates that the distance function is itself variable—evidence in disagreement with Anderson's assertion that distance remains constant regardless of the society occupying the space and is not subject to variations induced by social change. Carrothers's rebuttal was that in this type of analysis, distance, like time, only exists in the context of human interactance, and the physical facts of distance may well be meaningless as such (1956).

Seeking to overcome this difficulty, Carrothers (1956b), in a second study, utilized an income potential model (similar to the population potential developed by Stewart) in an attempt to explain interregional differences in population growth. Using income potential as "mass," the value of the distance exponent was found to be unity for western states, and the square for eastern and northeastern sections of the country. "This tended to substantiate the hypothesis that friction against human interactance, which is represented by physical distance between persons, tends to be higher in more densely developed areas." Pfouts (1958) also believed it to be desirable to study the function of distance as a variable. In fact, his concern about the aggregation of data in human interaction studies paralleled that of Carrothers, evidenced by his statement: "I am skeptical of any theories of social masses that cannot be related to individual motivations or actions" (Pfouts, 1958, p. 157).

As if in anticipation of this criticism, Carroll and Bevis

(1957) developed a local travel model based upon a modified gravity formula. This formulation assumed that individuals travel to satisfy a need or a desire. The distance traveled in this case is a function of the individual's ability to pay for travel (cost factor) and the population density (cohesive factor). All trips are composed of two basic parts—origin and destination—which are linked due to the needs or desires of the traveler. In an interesting approach, Carroll and Bevis envisioned land use as a two-way traffic generator—origin and destination. For "as people use the land, how they use the land determines how the people of the city move" (1957, p. 190).

The problem was then that of linking origin to destination. Three factors influence the choice of destination:

1. The probability of a trip's purpose being satisfied.
2. The accessibility of possible destinations.
3. The spatial competition between destinations (traffic, etc.).

Thus, this becomes a complex process in that the number of people traveling from a zone of origin to a zone of destination is a function of all possible destinations to choose between, and is in competition with trips originating in all other zones of origin desiring to travel to that zone of destination (1957). Note the implicit reference to Stouffer's theory of intervening opportunities.

Carroll and Bevis assumed zonal interchange volumes to be independent of each other when they examined deviations in the interchange volume reported in an origin and destination survey (Y_{ij}). By assuming that travel friction between all zones (X_{ij}) is of equal strength, the probability of interchange between any two points can be expressed as $X_{ij} = (Y_i Y_j)/Y$. The only determinants of zonal interchange in a frictionless space are the total number of trips generated at each zone (Y_i and Y_j) and the total volume of trips (Y). Using the ratio (Y_{ij}/X_{ij}), Carroll and Bevis were able to utilize regression analysis to estimate the relationship between travel time (their measure of friction) and zonal interaction. The resultant formula was:

(18)

$$Y_{ij}/X_{ij} = 12.68 D_{ij}^{-1.63}$$

where the constant and the distance exponent vary according to the trip purpose and destination. Unfortunately, Carroll and Bevis did not pursue their original intent of examining the effect of distance on the probability that a consumer will locate a destination which satisfies his needs. Nevertheless, their work did pave the way for more detailed investigations of the role of probability, distance, and human interaction.

Probability Approach of Morton Schneider

Perhaps the most important theoretical formulation to

emerge from the myriad number of empirical transportation studies was that of Schneider (1959), a research associate with the Chicago Area Transportation Study (CATS). Schneider presented a theoretical approach to the problem of trip distribution within a trip generating complex. Because he felt that all it provided was a specific answer to a narrow problem, Schneider was wary of gravity model's applicability; he was also very vocal in his criticism of empirical studies where the investigators were concerned in obtaining "best fit." He believed that "regression analysis, even log regression, is no more than a species, and a risky one, of curvefitting; it merely summarizes data" (Schneider, 1959, p. 52).

Schneider's major objections to utilizing the current gravity models, however, were that

> . . . There is no real kinship between a gravitational field and a trip-generating system. Newtonian gravity is an energy-force field characterizing the motions of particles, not their intentions. It does not deal in statistical partitions: a cluster of mass points under the influence of circumambient masses does not break into flights aimed at the separate attractive poles—each member of the cluster is content to be governed by the vector resultant obtaining at its point in space, and this vector varies from point to point only in mathematically continuous fashion. "Line of force" is a reasonable visualization, but this is not at all the same as lines of movement. Moreover, the computation of a gravitational force is uncompromising and unacceptable, the elements of the computation have stable dimension, meaning and measure. (Schneider, 1959, pp. 51-52)

Therefore, unable to accept that Newtonian laws of gravity were isomorphic with those of human interaction as proposed by the social physicists, Schneider sought explanation in a probabilistic framework. Schneider's construct provided for a probabilistic interpretation of a modified form of the intervening opportunity model developed by Stouffer, with an added constraint of a least-effort parameter not unlike that proposed by Zipf. Schneider's major assumptions were as follows:

> . . . The probability of a trip finding a terminal in any element of a region is proportional to the number of terminal opportunities contained in the element; that trip prefers to be as short as possible, lengthening only as it fails to find a terminal . . . the trip-receiving region is regarded as an unbounded plane surface over which trip, terminal opportunities are evenly distributed. (Schneider, 1959, pp. 52-53)

Within this conceptual framework, Schneider created a probability-density point function (G) whereby the probability (dP) that a trip originating at a point (O) will terminate within a region of area (dA) is (GdA). The probability (dP), however, of trip termination is the result of a joint probability—the probability (p) that a trip will reach (dA) and the probability (r) that it will terminate there once the zone has been reached. Therefore, $dP = GdA = pr$.

As the region is characterized by a constant trip-terminal density, (r) is assumed to be proportional to the area (dA). Also (p) is the probability that the trip will not terminate prior to reaching (O). Thus $dP = GdA$, and $G = sp$. Nevertheless, (p) is equal to one minus the probability that the trip finds a terminal, so

(19)

$$G = s(1 - \int_0^P dP)$$

or

(20)

$$G = s(1 - \int_0^A GdA) .$$

Schneider differentiates with respect to (A) and obtains

(21)

$$\frac{dG}{dA} = -sG .$$

Solving for (G), he obtains $G = ke^{-sA}$, where (k) is an integration constant equal to (s). Consequently, $G = se^{-sA}$ in the case of a plane surface $A = \pi t^2$, where (t) is the radial distance from the origin. Therefore

(22)

$$G = se^{-s\pi t^2} .$$

Schneider calculated the critical parameter (s) in terms of median trip length where

(23)

$$\int_0^{A_m} se^{-sA} dA = .5$$

and (Am) is the median area. Solving the equation reduces it to $s = -1/(A_m \ln .5)$. For the case where the destination zone is small enough to have a constant value for (G), the probability of a trip terminating in a given zone (j) of area (Aj) is:

(24)

$$P_j = GA_j = se^{-s\pi t^2} A_j .$$

Letting the number of trip-terminal opportunities in area (Aj) be (Vj), the following equation results:

(25)

$$P_j = s\frac{V_j}{\rho} e^{-s\pi t^2}; \rho = \frac{V}{A} .$$

By substitution we derive:

(26)

$$P_j = sA\frac{V_j}{V}\,e^{-s\pi t^2} \quad \text{or} \quad R\frac{V_j}{V}\,e^{-s\pi t^2} \ .$$

If the trip generation volume at origin (i) is (Vi), the expected interaction between origin (i) and destination zone (j) is:

(27)

$$V_{ij} = V_i P_j \ .$$

Schneider thus attempted to formulate a probabilistic foundation for human interaction within a spatial context through the creation of a probability density function. As with all modified gravity models utilizing the concept of intervening opportunities, selection of the extent of the spatial system is of critical importance. Regarding this problem, Schneider stated that "one must simply chose a large region and hope it is representative" (Schneider, 1959, p. 55). This question concerning spatial boundaries appears time and time again. Schneider, however, was able to introduce a probabilistic framework with a set of constraints based on the conservation of energy within his sytem—a least-effort approach to the spatial interaction of human groups.

Grecco and Breuning (1962) advocated new approaches to traffic flow analysis by stating:

> Although the techniques of systems analysis were developed primarily in electrical network analysis, during the past several years this fundamental discipline of analysis has been usefully applied to many other areas, such as mechanical, hydraulic, and heat-transfer systems. Prediction of traffic flows in an urban network also seems amenable to this technique. (1962, p. 12)

The implicit indication of the role of mechanics and energy levels (thermodynamics) within an urban system became important in the later works of A. G. Wilson, when he used the entropy concept to reformulate the gravity concept of human interaction.

Macroscopic Approaches to the Gravity Concept

In the midst of the flurry of empirical activity related to large-scale urban transportation studies, two figures emerged as major advocates of the continuing search for a more general spatial framework for social science. William Warntz and Walter Isard made substantial contributions to the evolution of the gravity concept of human interaction, despite the fact that neither conducted extensive empirical studies utilizing the gravity model per se.

William Warntz and Macro-Economic Geography

Warntz was critically concerned with bringing about a change in the prevalent microscopic approach of regional studies. He favored a more macroscopic framework. In order to avoid confusion in the interpretation of terms, Warntz stated:

> . . . The mere assembling of more and more areas, even with an increase in detail, does not mean a shift in point of view from microscopic to macroscopic. A heightening of the level of abstraction is the significant thing, an insistence of the functional consistency and organized unity of the whole, a recognition that no part of a true system can be thoroughly understood without reference to the whole. (Warntz, 1958a, p. 168)

With this approach, Warntz believed geography could become a truly theoretical and analytical science. The discovery of regularities in the aggregate would aid in forming a theory of human society. "The objective is to establish one social science and to show it and physical science are but mutually related isomorphic examples of one generalized logic" (Warntz, 1959a, p. 449).

Warntz utilized the irreducible dimensions of time, distance, and population in his search for a structure of spatial theory. Description was secondary, as Warntz's concentration was reserved for spatial patterns of aggregate phenomena. Warntz himself admitted that this concentration may require a "mechanical approach." Utilizing the concept of "field quantity," however, he believed it possible to explain the observed space-time-quantity regularities (Warntz, 1957a).

Field quantity theory was both a point of view (macroscopic) and a method of aggregate analysis. Warntz (1959b) felt that aggregate analysis was an integral part of the formative stages in the development of one generalized logic applicable to all knowledge. He believed this to be particularly true in geography where the treatment of space and distance was explicit. He explained it this way:

> The considerations of aggregate distributions in the spatial sense, that is, macro positions and aggregate access abilities, find expression in the mapping of the spatially continuous variables resulting from the analysis of the geographical distributions of "populations" considered as systems of integrated functional wholes. The varying levels of intensity can be derived only when the system is appreciated as constituting certain phenomena which are greater than the sum of the extension included. This point of view stands ready to permit economic geography to approach the level of science and overcome the difficulties associated with having only detailed microscopic examinations of discrete phenomena in nonanalytical distributions. (Warntz, 1958b, p. 59)

This macroscopic view of spatial phenomena was to provide the framework for all of Warntz's subsequent investigations.

Warntz was aware that empirical studies had uncovered a measurable factor of considerable importance: "sociological intensity." Its counterpart in physics was temperature, the level of molecular activity. Sociological intensity did not reside in a single individual but rather it was the effect of the combined influence of all people—an influence which varied geographically. Proximity of people to people was one of its major components. Nevertheless, it was not measured in terms of population density but in terms of a field quantity—potential of population (Warntz, 1959b).

The potential of population of any point in space is the summation of each individual's potential contribution (influence), which varies inversely with his distance from the point under consideration. Warntz assumed the potential of population to be a continuous function in space that could be represented by an integral of the form:

(28)

$$V_c = \int \frac{1}{r} DdA$$

where (D) is the density of population over an infinitesimal element of area (dA), and (r) is the distance. This formula is analogous to the gravitational potential of Lagrange. "To the social physicist, population potential is yet another aspect of the universal law of gravity. The geographer uses it to quantify the 'position' factor" (Warntz, 1956, p. 499). Population potential is in actuality a measurement of aggregate accessibility in which macropositions are quantified by means of integration, and in which the resultant coefficients of the spatial association are independent of a framework of a real subdivisions (Warntz, 1957a). Demographic energy became a unifying concept for Warntz in the investigation of human behavior.

Warntz's measure of social energy and its influence upon spatial interaction has intrigued investigators for some time, particularly those whose studies dealt with the gravity concept of human interaction. Voorhees (1955) believed that Warntz's concept of the potential of economic aggregates would enable researchers to establish more clearly which variables influence mass and distance factors in gravity models and their importance in spatial interaction. Investigators such as Stewart (1948) had previously demonstrated that geographic variations in a wide variety of sociological phenomena in the United States were highly correlated with spatial variations of potential of population. Warntz was firmly convinced that the utilization of the space potential technique would make it possible to measure the degree of a real association between certain kinds of geographical distributions in cases where such measures might not otherwise be possible.

Warntz was also concerned about the fact that geography, as human ecology, omitted entirely the consideration of the broad sociological aspects of spatial interrelatedness and interconnectedness of human activity. Physical distance, like time, is no less important a dimension in analysis of social phenomena than of physical phenomena. Thus, geographers have an opportunity to contribute to an integrated social science through the study of distance, not only as a social factor but also as a basic dimension in human ecology (Stewart and Warntz, 1958).

Need for Probabilistic Approach

Warntz also noted that "In regards to social science as well as in physical science, the behavior of the individual may not be considered as determined, but in both sciences aggregate behavior viewed macroscopically is determined and generalizations about it can be made, once the proper dimensions are isolated and recognized" (Warntz, 1958a, p. 45). Warntz was convinced that physics was the most organized and the most profoundly abstract science, and as such offered the best developed models for the search for the unifying principles within the social sciences. This accounts in part for Warntz's contention that

> . . . An emerging pattern of analysis in human geography combining both the macroscopic and the microscopic rejects the notion that is the crux of the possibilist-deterministic argument—i.e., that free will and determinism are incompatible. Here the microscopic refers to a decision-maker exercising his free will. The total environment . . . enters only as it is perceived by the decision-maker and included as a basis for his deliberation. Such an analysis is, of necessity, *ex ante*, microscopic, and probabilistic in nature.
>
> The macroscopic analysis attempts a consideration of actual milieu as it is related to on-the-average relationships among sociological phenomena and is thus *ex post* and deterministic, but only in the sense that it deals with tendencies toward spatial equilibrium. . . . Here cause and effect fade and the mutual relations of general patterns are stressed. (Warntz, 1959a, p. 435)

Within this frame, a macroscopic approach combined with a probabilistic interpretation at the microscopic level promised to provide geography with the means to make major contributions to the understanding of human interaction within a sociospatial construct.

These concepts were to influence profoundly the development of the gravity concept of human interaction. Warntz provided the means by which behavioralistic microscopic empirical studies could be incorporated within the deterministic framework of the gravity model. A probabilistic foundation for the gravity concept of human interaction became the desired goal for many of the social scientists who followed.

Warntz constantly sought to impress social scientists with the fact that

> . . . the spearhead of true progress is philosophical imagination and rationality complementing empirical observation . . . with this . . . man might realize the true essence of humanity, and no practicality could deter us from that necessary revolution in science, that next necessary great advance of knowledge, a truly natural science of society, giving proper emphasis to spatial as well as nonspatial processes and energy flows. (Warntz, 1967, pp. 18-19)

Warntz felt that in physics the profound attraction known as energy had supplemented space, time, and matter as an instrument of thought and, still more importantly, energy has been the concept which has integrated the diverse findings of many researchers. Cannot the same sort of reasoning be applied to the social sciences in which "social energy" performs the same function as physical energy in physics? Warntz, having posed the question, left the answer to future investigators.

Regional Applications of Walter Isard

A leading theorist during this time was Walter Isard. Although primarily concerned with the development of a viable regional science, Isard played an important role in the evolution and application of the gravity concept of human interaction. Isard's major concern was to improve the spatial and regional frameworks of the social science disciplines through the construction of a general theory of location and space economy. Opposing the traditional concept of regions, Isard argued that factors independent of the physical environment critically condition spatial configurations (Isard, 1956b). Yet he acknowledged that the "record of man's adaption to, and interaction with, his physical environment suggests that a comprehensive theory of society or economy should embrace both time and space dimensions" (Isard, 1956a, p. vii). The region for Isard became

> . . . a living organism, . . . whose spatial groupings of physical, biological and societal phenomena on the surface of the earth are functional in nature, and hence a proper field for analysis, . . . a whole complexedly interrelated with other regional entities and embodying internally an intricate network of interconnections. (Isard, 1956b, p. 17)

Nevertheless, a complete analysis of the living organism cannot be achieved simply by the summation of relevant studies in the existing social sciences. At the same time, Isard admitted that the present regional theories and analytic frameworks inadequately described the maze of relationships. The complexity of the regional interaction matrix resulted in a "characteristic common to, and in a sense forced upon, all forms of regional theories and analysis—the use of parts of the whole as conceptual or operating units and the lack of complete generality" (Isard, 1956b, p. 22). Isard felt, however, that slicing the region into various sectors provided theories of regional structure, however simple or whatever the size of the sector. Each theory or conceptual framework related to one or several sets of sectors (Isard, 1956b).

Search for the Agglomerative Force

A system of regions produced an even more intricate structure. Yet Isard was convinced that society was more than just a matrix of intricately detailed connections among units. Could not an overall force similar to agglomerative factors pervade society and confine the multitude of possible interactions among its innumerable units (Isard, 1960)? In attempting to explain this agglomerative force, Isard realized that his "fused framework" of regional analysis failed to

> . . . achieve a full perspective on agglomeration (spatial juxtaposition) forces and on the behavior of social masses. Such forces and behavior have not yet been successfully dissected by the efficiency approach of the comparative cost, industrial complex, and lines-programming techniques or by the requirements approach of input-output and other linear systems. The very size of the industrial agglomerations, urban-metropolitan masses, and social populations involved suggest exploration of the applicability of the probability-type gravity model. (Isard, 1960, p. 662)

In this case, from a purely theoretical standpoint, Isard found it not only possible but also fruitful and logically valid to embark upon abstract spatial theorizing. Interaction between regions becomes interaction between masses. Utilizing a set of uniformity assumptions, one may find it desirable to eliminate the physical environment of reality and to consider space only insofar as it entails resistance to movement, the so-called friction of distance (Isard, 1956). Thus Isard proposed utilizing the gravity concept of human interaction in the investigation of the differences in the frequency and intensity of spatial relationships.

Probability Approach to the Gravity Concept

Isard (1960) developed a gravity model which was based upon simplistic probability assumptions. Initially, a metropolitan region of population (P) is divided into many subareas. The total number of internal interactions (T) of the homogeneous population is known. The inhabitants interact in a frictionless world where travel costs are zero. Within this framework, the number of interactions which originate in subarea (i) and terminate in subarea (j) will *ceteris paribus* be equal to the ratio (P_j/P). The average number of trips per capita (T/P) equals a constant (k). In turn, the absolute number of trips an individual in subarea (i) makes to subarea (j) is $K(P_j/P)$. Since there are (P_i) individuals in subarea (i), however, the total number of trips (T_{ij}) from subarea (i) to subarea (j) is: $T_{ij} = k(P_iP_j/P)$. This represents Isard's probability postulate for interaction in a homogeneous frictionless world.

Aware, however, of previous empirical studies, Isard concluded that "it is undeniable that the friction of distance manifests itself in a number of important ways and markedly conditions the structure and functioning of critical sectors of the social system" (Isard, 1956a, p. 76). Isard looked for the effect of interactions between two subareas based on empirical data with the following technique. First, obtain the actual number of trips between every pair of subareas (I_{ij}). Divide the actual trip volume by the expected or hypothetical trip volume (T_{ij}) to obtain the ratio (I_{ij}/T_{ij}). By plotting the ratio (I_{ij}/T_{ij}) and the corresponding distance (d_{ij}) on a logarithmic scale, Isard found an apparent relationship between the log of the ratio (I_{ij}/T_{ij}) and (d_{ij}) with the resultant equation equal to:

(29)

$$\log \frac{I_{ij}}{T_{ij}} = a - b \log (d_{ij}) \ .$$

By eliminating the log f(x) from both sides and setting the antilog of (a) equal to (c), Isard obtained:

(30)

$$\frac{I_{ij}}{T_{ij}} = \frac{c}{d_{ij}^{b}} \qquad \text{or} \qquad I_{ij} = c\frac{T_{ij}}{d_{ij}^{b}} \ .$$

With the substitution of G — ck/P and a previous term for (T_{ij}), a classic form of the gravity model emerges as:

(31)

$$I_{ij} = G\frac{P_iP_j}{d_{ij}^{G}} \ .$$

Isard regarded this concept as a basic principle underlying the structure of metropolitan regions and systems of metropolitan regions (Isard, 1960).

Problems of Disaggregation

Previous investigators had, however, disaggregated the model into a more complex form such as

(32)

$$I_{ij} - \frac{Gw_i(P_i)^{\alpha} \cdot w_j(P_j)^{\beta}}{d_{ij}^{b}} \ .$$

Isard argued that disaggregation, when it eliminated a meaningful aggregate, precluded the otherwise reasonable assumption that individual ingroup irregularities are destroyed. In other words,

> . . . disaggregation is desirable when additional information and precision is obtainable and when such disaggregation does not destroy to any great degree the inherent meaning and internal structural unity of the mass or population. Under these circumstances, it will be fruitful to employ distinguishing exponents, weights, etc. (Isard, 1960, p. 515)

Unfortunately, this warning went unheeded in much of the research that followed its issuance.

Social Distance

Isard realized that current gravity models lacked any theory which could explain the values of those f(x)'s assigned to weights and exponents. Isard called for

> . . . an imaginative and yet vigorous inquiry into the theoretical foundations of the gravity models. This inquiry would certainly explore thoroughly the probability basis for the gravity model. It would necessarily investigate whether or not an optimization process is involved, and the relation of the gravity model to the rank-size rule and to agglomeration and spatial juxtaposition economies. . . . This inquiry . . . might examine the ability of the concept of intervening opportunities to represent various forms of *social distance,* and hence to be of basic significance for gravity model investigation into the behavior of social masses. (Isard, 1960, pp. 751-52)

Aware that a precise definition of "social distance" had not yet been achieved, Isard suggested that such factors as level and type of information possessed by interacting units (later to be used by Huff), the binding force of cultural patterns, and linkages among social and economic roles be carefully analyzed. Fortunately, much of the analysis developed for the concept of physical distance was valid in its parallel application to economic distance because of similar characteristics. Social distance is subject to change from institutional and cultural developments, just as physical and economic distance are influenced by advancements in technology and transportation. Because social distance lacked a precise definition, however, Isard preferred to view the gravity model as being applicable only to whole integral masses, although he did admit that the concept of intervening opportunity might provide valuable insight into the special characteristics and patterns of any given sector of the regional mass (Isard, 1960).

Commenting on the efforts of Zipf and others to link the gravity model to the rank-size rule, Isard pointed out that "implied in such linkage is an optimization process based on some crude probability reasoning. If such an optimization process is in fact involved, as intuition strongly suggests, one's understanding of the gravity model will be significantly advanced by an explicit statement of the process" (Isard, 1960, p. 564). This optimization process was to be investigated by Huff (1959, 1960) under the classical economic concept of "utility."

Finally, Isard appealed to future researchers to

> . . . develop new and superior concepts relating to the spatial structure of society. Spatial interaction phenomena, as manifested for example in the various empirical materials on commodity and communication flows and population movement, must be dissected with tools honed to a much finer sharpness. We especially need to probe into *space preferences,* i.e., into man's propensity for intricate forms and patterns of herd existence and into the socio-psychological and biological forces which together with economic and other forces govern the spatial patterns of population settlement. (Isard, 1960, p. 287)

This would provide greater insight into the structure and functioning of metropolitan regions as socioeconomic organisms.

Space Preference

One possible avenue of exploration recommended by

Isard was analysis of the individual person rather than distinct production processes. The space preference of the individual (that is, his propensity or level of social interaction) would be examined. This essentially behavioralistic approach was to be ideally suited to those geographers with theoretical inclinations who started to probe causal analysis in a multifactorial manner, more characteristic of economics and sociology than classical geography (Isard, 1960).

Isard's major concern was with the development of a regional science which involved the power of abstractions within the pure social, political, and economic frameworks of that science. His timely analysis of the conceptual and methodological requirements of the gravity concept of human interaction provided needed direction for future research. Isard's visionary abilities enabled him to forecast the most promising approaches in evolving a general purpose regional structure based upon a general theory that would capture the essence of a region as a dynamic organism (Isard, 1956b). His constant voice imploring investigators to "strive for more appropriate conceptual frameworks, venture new hypotheses and models of interrelations, explore new arrangements and processing of data and otherwise thrust forward with vigor" (Isard, 1956b, p. 13) provided critical momentum to the evolution of the gravity concept of human interaction. Both Warntz and Isard stand out during a period generally characterized by numerous microscopic empirical studies of human interaction because of their continual pleas for a more macroscopic approach in the formulation of concepts of social interaction.

IV. SEARCH FOR A THEORETICAL FOUNDATION FOR THE GRAVITY CONCEPT

Despite the plethora of empirical studies utilizing the gravity concept of human interaction, many social scientists continued to maintain that the gravity theory was essentially an empirical tool. In their minds, the postulate lacked a firm theoretical foundation and was incapable of explaining observed regularities (Huff, 1961). Previous studies, from overly simple empirical rules (Reilly, 1961) to complex formulations (Nystuen, 1959; and Garrison, 1959), had described the patterns of human interaction. The differences in human propensity for spatial interaction, however, had yet to be fully understood. The deterministic nature of the gravity model was rejected. Reformulation was necessary. Investigators sought stochastic interpretations of the classical gravity concept through the application of theories borrowed from behavioral psychology and economics. One such investigator was Huff.

Behavioral Approach of David Huff

Huff vigorously opposed the traditional theories of consumer travel which stated that: "the major determinant of spatial interaction is that of proximity . . . trip frequency as well as total distance travelled declines with increasing distance . . . the forces underlying individual movement are those that are inherent in the spatial system itself" (Huff, 1959, p. 27). Huff maintained that the daily patterns of human interaction were greatly influenced by the spatial structuring of society (Huff, 1960). Yet Huff also believed that there were forces outside the spatial system which determine or influence human interaction. An empirical study conducted by Stone and Form (1957) indicates that variations in consumer shopping behavior were a result of differences in socioeconomic characteristics. Their finding was supported shortly thereafter, when Marble (1959) discovered that there was a low correlation between total distance traveled and variations in residential location. There also appeared to be a significant relationship between the socioeconomic status of an individual and the total number of trips made by him, as well as the distance traveled.

Decision-Making Process

It was apparent to Huff that any systematic approach to human spatial interaction would require a broader conceptual framework than that of the classic gravity model if exogenous socioeconomic forces were to be correctly analyzed. Dodd had previously attempted to introduce these variables as multipliers of his basic variables; however, his formulation

(33)

$$F_{ij} = k \frac{\Sigma \phi P_i \Sigma \psi P_i}{D_{ij}} .$$

had not been refined to the point that it could be used to accurately predict spatial behavior within urban areas (Huff, 1962b). Huff was dismayed at how little had been done in the way of examining the behavioral implications associated with the usage of the gravity concept (Huff, 1963). Therefore, he sought to analyze the decision-making process by which goals and intended destinations are chosen by consumers. This decision to construct a theoretical abstraction of consumers' spatial behavior (Huff, 1964) would gradually evolve into one of the best known examples of a modified gravity model of human interaction.

Aware that "a general theoretical scheme for analyzing individual travel movements must consider the behavioral characteristics of a person in addition to and relative to his environment," Huff examined theories current in behavioral psychology. He sought a descriptive, graphic illustration which was capable of explaining the relationship between various socioeconomic and spatial factors influencing consumer movement. The vector-valence concept was selected.

Vector-Valence Concept

Originally developed by Lewin (1958) in order to explain

human decision-making, the vector-valence concept was instrumental in forming the basis for Huff's initial formulations of spatial behavior. In its most basic form, the vector-valence concept explained that human decision-making was a result of an internal conflict in which an actor weights the intensity of his desires against the intensity of his resistance. The actor assigns a different valence to each of these positive and negative forces—the resultant action is dependent upon the net value attached to each alternative.

Huff's application of the vector-valence concept assumes that a consumer's desire for goal attainment (goods or services) *as well as its locational source* depends upon the consumer's net evaluations of all socioeconomic and spatial factors influencing the goal attainment. In this context, the locational source chosen is the one that displays the greatest net positive-valence (Huff, 1959). Thus it is possible, under this construct, for a consumer to patronize a store more distant than one offering identical goods and services—a direct contradiction of an accepted postulate of consumer theory. Huff implied that the vector-valence concept could account for the findings of Marble's (1959) study of consumer behavior, which concluded that there existed a low correlation between total distance traveled and variations in residential location within an urban area. Marble showed that trip frequency was affected by a consumer's residence relative to the retail structure of the city.

Marble's findings did not lead to the complete abandonment of a distance variable by Huff. Nevertheless, distance represented only one of many factors in the decision-making process and was not assigned the dominant position, as it had been by previous theorists using the gravity concept. Huff contended that the net value curve of intraurban travel did not appear to be a continuous function of distance traveled by individual consumers. The fact that there was an implied range of travel for certain product classes, which on the average was representative of the entire population, was acknowledged by Huff. He was concerned, however, that this phenomenon did not explain the differences exhibited by different social groups relative to their spatial activities.

Use of Isard's Space Preference Principle

Huff apparently was influenced by the works of Isard (1956) concerning the development of general theories of location and space-economy. In searching for variants of currently accepted approaches to the explanation of spatial flow phenomena, Isard suggested that the analysis might start first with individual persons rather than with individual production processes. With this "space preference" concept, Isard felt that

> . . . It may be necessary to enter the realms of sociology, and social psychology in order to explain the spatial distribution of household consumers around focal points—for example, the population spread around any given metropolitan

> core—for this requires knowledge of the process by which tastes are molded and, in particular, understanding of the spatial preferences of consumers. . . . Psychologists and sociologists, whether speaking of gregarious instinct or of acquired behavior patterns or both, have emphasized the social nature of man and his propensity to associate with groups of various sorts. One can reason that such a propensity, acquired or instinctive, is a manifestation of a positive space preference. . . . Thus the spatial preference of each person, both as a consumer and as an income producing unit, would be considered. Aggregating individuals to form meaningful social and economic groups would introduce groups space preference. . . . Human ecology promises eventually to provide such an understanding. (Isard, 1956, pp. 22-23, 84-85, 144-45)

This concept of individual space preference provided Huff with the needed direction for his study of human interaction. He saw that individual space preferences determined by sociopsychological and economic forces varied among individuals and that these variations existing within the social nature of man affected his propensity to interact with various groups.

Individualistic Behavioral Approach

As a result, Huff, in contrast to previous attempts at formulations of aggregate human spatial interaction, sought explanation for individual spatial activity (Huff, 1960), with his basic assumptions founded upon behavioralistic assumptions. Each consumer was seen as being faced with a stimulus situation whereby his interactions were constantly being affected by physical, social, and cultural objects. In turn, these objects, combined with the consumers' psychological drive, produce a need for some type of consumptive behavior, if there is a readiness on the part of the consumer to seek the object which will satisfy his need—"desideratum."

It is at this point in the decision process that various socioeconomic variables influence a consumer's decision to seek satisfaction. A consumer's value system will condition his perceptions and actions directed toward goal satisfaction. Therefore, as the variety of goal alternatives with their *resultant locations* is scanned, the net value attached to each depends upon the socioeconomic characteristics of the consumer's value system. Huff lists "geographical location, ethical and moral code, ethnic affiliation, income, sex, personality, occupation, age, education, and mental synthesizing abilities" (Huff, 1960, p. 62) as probable value system variables.

The actual selection of the locational source which will satisfy a consumer's needs is influenced by such factors as: reputation of the source, personal amenities, breadth of merchandise, services rendered, and price of product. These factors are referred to as the "behavior-space perceptions" by Huff and are similar to what others call the "attractiveness" of a center. In addition, elements of "movement imagery," which reflect difficulty in reaching goal satisfaction sources, can be expressed in terms of transport mode, travel time, and travel and parking costs, as they influence the selection of the final destination. A restructuring process is also at work, once the locational source has been visited, on the consumer's perception of the center as well as on his retention of the pattern of movement required to reach the locational source. The exact effect of the restructuring process, especially in its effect on formation of habitual patterns of shopping behavior, has yet to be explored.

Despite the seeming complexity of Huff's decision-making process as a basis for the explanation of human interaction, it is possible to relate it to a simple variation of the classic gravity model. Huff's model is, in fact, compatible with the gravity concept if it is restated as follows:

(34)

$$\text{Overt Behavior (Interaction)} = \frac{\text{(Value System)(Desideratum)(Behavior-Space Perception)}}{\text{(Movement Imagery)}};$$

a behavioralist formulation of one of Dodd's basic gravity concepts where

(35)

$$I_{ij} = \frac{GU_i N_j}{d_{ij}} \; .$$

(G) is replaced by "desideratum," the force of inertia or gravity; (U_i) by the molecular weight attached to the socioeconomic class of the individual; (N_j) by the force of "attraction"; and (d_{ij}) by the "movement image" or cost of movement.

The basic conceptual difference between Huff's behavioralistic formulation and the gravity concepts of Stewart and Dodd was that Huff described spatial interaction in terms of individuals, not aggregates of individuals. Huff's formulation provided a behavioralistic framework for obtaining insight into those socioeconomic spatial variables which influence human interaction or space preferences, but his model was only the first stage in Huff's development of his well-known probabilistic model of consumer behavior. Aware of the difficulties in predicting individual spatial behavior adequately, Huff investigated the possibility of grouping individuals according to various similarities in order to "predict the types of locational sources consumers of various sorts will choose for different goods and services as well as the distances and frequencies of travel related to this type of activity" (Huff, 1960, p. 173). With this objective in mind, Huff reexamined the gravity concept of human interaction.

The existing gravity model in terms of consumer behavior had the form:

(36)

$$F_{ij} = a \frac{A_{kj}}{D_{ij}^{b}} ,$$

where

F_{ij} = expected frequency of interaction between point (i) and destination (j).

A_{kj} = attraction of the (j)th destination.

D_{ij} = distance from point of origination (i) to the (j)th destination.

a = a constant of proportionality.

b = a constant exponent.

This model stated that the interaction between a consumer and various locational sources varies directly with the attraction of the source and inversely with the distance to that source. Huff (1962b) was quick to point out major limitations of this formulation as follows:

1. Because gravity concept utilized the aggregate behavior of people in order to predict group behavior on the basis of the laws of probability, it did not attempt to explain individual interaction.

2. The gravity concept was unable to account for variations in the spatial interaction of nonhomogeneous groups or between various socioeconomic groups.

3. The distance exponent was raised to some power greater than unity; its value was determined by the trip purpose despite evidence presented by various investigators that the distance exponent may be a function of distance itself, socioeconomic variables, or density.

4. The gravity model, deterministic in nature and essentially a static formulation, was unable to forecast future interaction.

5. The gravity model, deterministic in nature and basically an empirical notion, lacked a theoretical foundation.

Huff viewed these models as tools which allowed short-run approximations of the direction and magnitude of individual travel patterns. As such, it appeared unrealistic to classify these models as reflecting certain general laws of human interaction (Huff, 1962b). Nevertheless, Huff noted that the gravity concept contained one basic principle of human interaction: the principle that interaction declines with distance. In an attempt to circumvent the inherent limitations of the gravity model, Huff returned to a conceptual analysis of consumer spatial behavior. What Huff recognized was the possibility of utilizing certain conceptual properties of the gravity theory—in fact, a conceptual reformulation of the gravity model of human interaction. Heeding Isard's suggestion, Huff analyzed the ecological characteristics of consumer behavior (Huff, 1961, 1962). In the course of these studies, Huff concluded that five factors were the major explanatory variables relative to spatial patterns of consumer behavior. These factors were: (1) merchandise offerings, (2) travel costs, (3) product classes, (4) consumer income, and (5) city size, or density of population.

Probabilistic Assumptions

Huff introduced the first of a series of probabilistic assumptions in the process of formulating his model of consumer behavior.

> . . . A large number of consumer shopping decisions are made under conditions of uncertainty, i.e., the consumer does not know in advance whether a particular locational source will necessarily fulfill a specified purchase desire. However, the consumer does have *a priori* knowledge of the probability that various sources might satisfy his shopping demands. Such a probability is based for the most part, on the number of items of the kind that he desires or feels are carried by each of the various locational sources. Therefore, consumers will show a willingness to travel further distances for various goods and services as the number of such items available at various locational sources increases. (Huff, 1961, pp. 20-21)

Consumer spatial behavior was seen to be significantly influenced by the time, effort, and expense perceived to be involved in traveling to the locational source. Note that actual physical distance used in the gravity model is replaced by "social distance" similar to Thompson's (1963) concept of "subjective" distance. The function of total travel cost, increasing with distance at an increasing rate, tends to limit the distance traveled for any particular good or service (Huff, 1963). The limit or range of a good appears to be a function of the good's relative price, frequency of purchase, and degree of product substitutability. "Therefore, in any analysis of consumer travel behavior, one can think of a 'family' of product curves which will vary depending on the relative values that such products possess for different consumers" (Huff, 1961, p. 25).

In terms of economic variables, Huff argued that income differences were a major factor in explaining why consumers from higher economic groups tend to travel greater distances for similar shopping purposes. Unfortunately, Huff was unable to determine if this phenomenon was actually a result of a lower relative travel cost with respect to income or a result of the fact that residential zones of higher socioeconomic classes tend to exhibit a more decentralized retail structure and lower population density.

This problem, however, had been addressed earlier by Carrothers who stated that

> . . . Friction per unit space against interaction caused by short distances is disproportionately greater than friction per unit of distance caused by longer distances. For instance, friction against movement within an urban area is generally greater than that caused by an equal distance in the less densely populated space between two such areas. Or, again, an extra unit of distance added to a long movement is less important than an extra unit added to a short movement. (Carrothers, 1956, p. 97)

Stouffer (1960) added the effect of competing consumers

for the same locational source as a contributor to the value of friction in intraurban travel.

With ecological variables identified, Huff realized that a model containing many complex independent variables would result in an endless number of relationships and would be difficult to analyze. He therefore concentrated on two variables: shopping center attractiveness and consumer travel costs. This enabled him to return to the two major components of the classic gravity concept: mass (attraction) and distance (cost). Huff, however, attempted to provide a theoretical base for his selection of these two factors—the economic theory of utility.

Utility Theory

Within Huff's construct, the consumer is faced with determining which location yields the best balance between benefit and cost. This relationship determines the maximum limit that a consumer is willing to travel for any particular product or service. In order for interaction to take place between a consumer at point (i) and locational source (j), it is necessary that $V_{ij} > C_{ij}$ and $V_{ij+1} - C_{ij+1} > V_{ij} - C_{ij+1}$, where ($V_{ij}$) is the value of the "payoff" or the utility of moving from (i) to (j), (C_{ij}) is the cost incurred moving from (i) to (j), and (V_{ij}) is the incremental value obtained by moving to a more distant location (Huff, 1963c). Perhaps the expression of this relationship is Huff's greatest contribution not only to consumer behavior theory but to the gravity concept of human interaction as well. The fact that the existence of centers of population or retailing does not ipso facto result in interaction is due to simultaneous decision processes which determine the net utility of spatial movement to all possible goal-satisfying locational sources. The *utility function* becomes the source of energy that initiates human interaction. When a consumer is faced with a set of alternative locational sources, analysis of spatial behavior increases in complexity. Nevertheless, Huff incorporated a probabilistic interpretation in his behavioralistic assumptions concerning locational source utility.

Huff's Model of Consumer Behavior

Within this probabilistic framework, the basic elements of Huff's model of consumer spatial behavior are:

1. The set of shopping center choices represented by set (J).

2. The alternative set of shopping center choices represented by set (J_o). This subset of set (J) conforms to the tastes and preferences of consumers. Any given alternative is written as (j), where (j = 1, 2, 3, . . . , n).

3. The positive "payoff" function (uj) associated with the utility of each alternative shopping center.

4. The shopping center size (S_j).

5. The empirically estimated parameter (λ) which reflects

the effects of travel time on various types of shopping trips.

6. The travel time (T_{ij}) required by consumers to get from base (i) to shopping center (j), (Huff, 1963c).

The major postulates of Huff's theory of consumer spatial behavior can be stated as follows:

1. The probability (P_j) of a given location (j) being chosen among all other alternative locations in the subset (T_O) is proportional to (uj), where

(37)

$$P_j = \frac{\mu_j}{\sum_{j=1}^{n} \mu_j} ,$$

if

(38)

$$\sum_{j=1}^{n} P_j = 1 \qquad \text{and} \qquad 0 < P_j < 1 .$$

2. The ratio between the probabilities of a consumer's choosing one of two particular shopping centers does not depend on the existence of other centers. This ratio is referred to as the ratio of the utilities of the two centers to a consumer. Therefore, $P_{j1}/P_{j2} = u_{j1}/u_{j2}$.

3. The properties of the pair (P_{j1}, P_{j2}) that determine the utility (u_{j1}, u_{j2}) are size (S_j) of the shopping centers, and distance (T_{ij}).

These postulates combine to form Huff's probabilistic model of consumer behavior where:

(39)

$$P_{ij} = \frac{\mu_{ij}}{\sum_{j=1}^{n} \mu_{ij}} = \frac{\dfrac{S_j}{T_{ij}^{\lambda}}}{\sum_{j=1}^{n} \dfrac{S_j}{T_{ij}^{\lambda}}} .$$

Probabilistic formulation differs from traditional consumer theory, which had stated that consumers will choose one particular alternative (the most desirable one) with a probability of unity, all other alternatives being zero. In contrast, Huff asserted that consumers are not able to distinguish among alternatives perfectly and thus are not capable of maximizing their utility in the sense of choosing one alternative exclusively. Huff based this assumption on two factors:

1. When differences in utility are small, consumers choose alternative centers in a somewhat random pattern.

2. If the consumer is uncertain of the "payoff" associated with each center, then he will select one from among all the shopping centers. At this stage, the consumer, through the behavioral decision-making process previously described, utilizes an intuitive evaluation of each

> center. If this process is repeated, the consumer will select alternative locational sources in some constant ratio based upon the "relative utility" of each center. (Huff, 1962a)

These assumptions account for the probabilistic nature of Huff's model of consumer spatial behavior. Within its basic conceptual construct, however, the model is behavioralistic. Unfortunately, this fact is generally overlooked, because Huff, possibly out of necessity, chose store size and travel time as his major explanatory variables. Huff's model thus emerges as a composite formulation. The numerator and denominator are in the classic gravity form, yet the entire formula is a modified expression of Stouffer's model of intervening opportunities. Not surprisingly, difficulties inherent to both types of models plagued Huff.

Lewis and Traill (1968) noted that Huff's assumption that the exponential value of the attractiveness index is unity was not a reflection of reality. They preferred to include a variable exponent "which adds something to the attractive forces of each single shop when compared with one that stands alone" (Lewis and Traill, 1968, p. 321). This variable exponent reflected the force of agglomeration upon consumer behavior. The weakness of the model is Huff's failure to provide a basis upon which the extent of the closed system could be determined. Areal boundaries must include all potential centers for which the consumer's probability of visiting is greater than zero, so that summation of all these probabilities can equal unity:

(40)

$$\sum_{j=1}^{n} P_j = 1 \ .$$

Since the size of the consumer's shopping zone determines the number (n) of alternative centers available to him, it is of critical importance in determining the probability of interaction. The problem becomes more complex in intraurban regions where centers are numerous and vary greatly in terms of size and spacing, that is, from neighborhood centers to the central business district. Regardless of these imperfections, Huff's model has found widespread acceptance in marketing research, particularly in trade area studies. Yet Huff's greatest contribution to the gravity concept of human interaction does not appear to be his model of consumer spatial behavior; rather, it is his attempt to provide a theoretical foundation for spatial interaction based upon valid behavioralistic postulates amenable to probabilistic interpretations. Thus, through Huff's many stages of model formation emerges a construct that bridges the gap between the macroscopic theories of the "social physicists" and the microscopic approach of the behavioral psychologists.

Huff's work contributed substantially to a fuller understanding of the nature of "social or subjective distance" and its conceptual usage in the gravity concept of human interaction.

Subsequent Behavioral Studies

The spatial analysis of consumer behavior became the topic of investigation for a great number of researchers. Thompson (1966) focused on the validity of the assumption that distance constitutes a friction or inertia which tends to keep retail activities localized in any given area, and on the assumption that *actual* distance corresponded with the consumer's *estimate* of distance, convenience, and travel time.

Thompson's Subjective Distance

A general tendency on the part of consumers to overestimate both distance and travel time was uncovered. Thompson, in analyzing the ratio of the overestimation, concluded that the degree of overestimation was directly related to the shopper's attitude toward the center in question as well as the level of merchandise offerings and convenience. As the study indicated that the perceived distance differed from the actual distance, Thompson proposed that "objectively determined distance and driving time measures may not be entirely appropriate inputs for simple models designed to describe or explain geographic patterns of consumer purchasing behavior" (Thompson, 1963, p. 6).

Thompson questioned the consumer choice process implicit in Huff's formulation of consumer behavior because it appeared to be at variance with the observations of other market researchers. Investigators such as Katona (1960) and Alderson (1957) contended that habit and inertia are the relevant dimensions of consumer behavior; they found that problem solving was a rare occurrence.

Thompson was also critical of Huff's arbitrary selection of a twenty-mile radius as including all perceived alternative centers of the individual consumer. Thompson felt this process reduced a critical variable in the decision-making process to a constant. In effect, a vital variable—that is, the less than perfect ability of the consumer to perceive the urban or suburban environment in all its complexities—was negated (Thompson, 1966). Thompson contended that any model of consumer behavior should calculate utilities only for those portions of the retail landscape effectively within a consumer's problem space. Thompson suggested that a model of consumer spatial behavior should be based on the subjective landscape interpretations of consumers who travel it—*the perceived urban structure.* Thompson was aware of the interpretive difficulties inherent in this behavioral approach. With the complexity of the urban structure evident, Thompson acknowledged the possibility of developing an objectively defined consumer decision-making model of unmanageable proportions. Thompson was encouraged, however, by the findings of a study conducted by Lynch (1960), which attempted empirical determination of an individual's reaction to the complexity of the urban environment.

Lynch's Perceptual Approach to Distance

Lynch was able to show that there were indeed significant differences between the dimensions of the city as it existed in space and the dimensions of the city as it existed in the minds of the inhabitants. Most important, however, was the discovery that there are observable regularities in the manner in which subjects tend to organize their perception of the urban structure. Images of the city tend to be based on key landmarks, places of residence, and places of employment—the movement corridors of the individual.

Thompson recommended continued research of an individual consumer's subjective orientation to the complexity of the urban environment as one of the more promising arenas of research for seeking explanation of consumer behavior. The imagery concept would focus on the individual's image of the retail structure relative to the entire urban form. Thompson encouraged concentrated effort to determine if image difference can be explained in terms of measurable consumer attributes such as age, length of residence, education, income, race, social class, and so forth. He was also interested in how the image developed over time and the process of image modification. While Thompson's imagery concept differs from the problem-solving, decision-making method used by Huff, it must be noted that his initial construct based a center's utility on consumers' perception of that center.

The Work of Transactional Psychologists

Another possible conceptual framework for the explanation of social or subjective distance has been explored by "transactional" psychologists such as Kilpatrick (1961). Transactional psychologists view individual perception as a process involving interdependent subjective influences which act upon an individual's awareness of the world about him. This awareness emerges from a complicated weighing process where the nature of the stimulus of the external environment is combined with past experiences of similar situations. Thompson was convinced that

> . . . The study of individual approaches to the interpretation and organization of complex phenomena in experimental situations, and the reaction to problem solving may well be of potential value in the formulation of more meaningful theories of shopping behavior . . . and it may be argued that the transactional psychologists embrace a conceptual framework which is of direct relevance to the analysis of the behavioral base of shopping behavior. (Thompson, 1966, pp. 14-15)

Thompson's studies have introduced an interesting dualism into the gravity concept of human interaction. Empirical studies have indicated that an individual's subjective bias and perception of the urban environment affects his ability to judge objective magnitudes. Should the empiricist abandon his preoccupation with seeking objective variables (distance, travel cost, and so forth) and focus on subjective individual perceptions as a basis for formulating the gravity concept of human interaction? In any event, it is evident that a more interdisciplinary approach is needed for a more complete understanding of human spatial interaction, whether or not it is based upon behavioral assumptions.

Opportunity-Claimant Model of Lewis and Traill

A pair of investigators, Lewis and Traill (1968), were also involved in the analysis of consumer behavior. Making use of Huff's basic formulation, they transposed it to the form:

(41)

$$Q_{ij} = \frac{\frac{A_j^\alpha}{d_{ij}^\alpha}}{\sum \frac{A_j^\alpha}{d_{ij}^\alpha}} ,$$

where (Q_{ij}) is the proportion of total spending of residents of zone (i) which takes place in zone (j). In contrast to Huff's formulation, an exponent is used in conjunction with the attractiveness of a center (A) to account for the added attractiveness of the center due to forces of agglomeration. Lewis and Traill held that the attractiveness of a grouping of similar stores or services was greater than their sum—a concept presented much earlier by Walter Isard. The most important aspect of Lewis and Traill's model (in the sense that the model differed from that of Huff), however, was the introduction of an "opportunity-claimant" factor. The basic postulates were developed in the following manner:

> . . . There are opportunities (to shop) which are distributed between zones. . . . There are also claimants for these opportunities, living in various zones . . . this second kind of competition has been ignored . . . at any given location of opportunity there may be competition between the claimants. The importance of the competition clearly varies from activity to activity, and in some cases it may be difficult to decide where it begins . . . claimants for these opportunities in zone (j) have different degrees of effectiveness, and . . . this depends on the distance of their location from (j). (Lewis and Traill, 1968, p. 322)

Thus the total number of effective competitors for the opportunities in zone (j) is

(42)

$$C_j = P_j + \sum_i \frac{p_i}{d_{ij}^\beta} ,$$

where (p) is the population of claimants located in the zone, (d) is distance, and (b) a distance exponent. If the attractiveness of any zone (j) depends on both the volume of opportunities (A_j) and the volume of competing consumers, the attractiveness of zone (j) is written as:

(43)

$$\frac{A_j^{\alpha}}{C_j^{\delta} d_{ij}^{\lambda}} ,$$

and the entire expression can be represented as:

(44)

$$Q_{ij} = \frac{\dfrac{A_j^{\alpha}}{C_j^{\delta} d_{ij}^{\lambda}}}{\Sigma C_j^{\delta} d_{ij}^{\lambda}} .$$

What Lewis and Traill did was to conceptually and mathematically formulate a parameter of social friction. The relationship between urban density and social friction is in need of further investigation.

Lewis and Traill, however, introduced yet another element into the already complex question of social or subjective distance—that of social friction. Another author, Golledge (1970), also noted the current trend to include realistic behavioral assumptions in models of consumer behavior. To Golledge, there appeared to be a continued emphasis on dynamic equilibrium models, which have short-run solutions but yet allow for behavioral change as a result of internal changes of sample members or as a result of changes in the environment. One result of this emphasis has been the fusion of economic spatial and behavioral theories of equilibrium in an effort to formulate general theories of human spatial interaction.

Present State of Behavioral Models

After examining the the literature on behavioral models capable of producing spatial equilibrium outputs (relative to consumer behavior), Golledge concluded that

> . . . most equilibrium models can only be expected to explain a small part of the actual behavior of population. Herein lies a challenge for the geographer—to determine the spatial patterns of different types of behavior to build appropriate models, and to state explicitly what proportions of a randomly selected sampled population's behavior can be explained by the model. It is quite feasible that, to achieve any large degree of explanation, more attention may have to be focused on nonequilibrium models, of both spatial behavior and spatial systems. (Golledge, 1970, p. 423)

Golledge has succintly summed up present attempts to provide a behavioral base for the gravity concept of human interaction. Three areas appear to require further research: the role of perception of the urban structure, the process of decision-making which initiates the interaction, and the role of repetitive movement along continuous corridors of travel. These are the three obvious spatial factors of human interaction—origin, destination, and linkage—but viewed within a subjective context.

Composite Formulations Based on Theories from Economics

The Work of Niedercorn and Bechdolt

Closely related to these behavioral studies have been complicated theories derived from economics, theories explaining the rational man. One of the most influential works written during the "composite" period of the gravity concept's evolution was that of Niedercorn and Bechdolt (1969). This detailed study derived a "gravity law" of spatial interaction within the framework of utility theory. The authors identified three primary groups of factors which affect the spatial interaction of social phenomena:

1. The origin factors—elements of an origin which indicate its ability to interact with all possible terminal points.
2. The destination factors—characteristics which explain the relative volume of interaction with all others—attractiveness index.
3. The linkage factors—relationships between the origin and the destination.

Expressed mathematically, spatial interactions are functions of one or more of the origin, destination, or linkage variables.

Niedercorn and Bechdolt's derivation is given in terms of a single origin/multiple-destination model where the basic assumptions are as follows: given a subnational region composed of (n+1) areas, the object of the study is the spatial interaction behavior of an individual (k), at origin area (i), (i = 0) with each destination (j), (j = 1, 2, 3, . . . , n). Assume that there is only one locational goal at each destination (j). Therefore, the individual's utility of the trip from (i) to (j) is a first approximation of the individual's utility of interaction with (j), which can be expressed as $kU'_{ij}f(kT_{ij})$, where kU'_{ij} equals the net utility of individual (k) at origin (i) interacting with destination (j) per unit time, and $k\ T_{ij}$ equals the number of trips of (k) from (i) to (j) per unit time. The individual's total net utility (kU_i) with all destinations is:

(45)

$$\sum_{j=1}^{n} f(kT_{ij}) \ .$$

Niedercorn and Bechdolt assumed that the number of opportunities or locational goals in any zone is proportional to the population of each zone. With the restriction that an individual can only make one interaction per trip, the individual's utility function of interaction with all opportunities in all zones is:

(46)

$$kU_i = a \sum_{j=1}^{n} P_j + f(kT_{ij}) \ ,$$

where (KU_i) is total net utility, (P_j) is population at (j), and (a) is a constant of proportionality.

Niedercorn and Bechdolt introduced two constraints on consumer travel study, often ignored by other investigators.

A consumer cannot take an unlimited number of trips, due to his own travel limits. He is constrained by:

1. The total amount of money available to him for travel expenses, expressed as $k(m_i)$.

2. The total amount of time which he allocates for travel, $k(H_i)$.

The equation for (T_{ij}) is similar to the "gravity law" where (T_{ij}) is inversely proportional to (d_{ij}) raised to the first power.

The second trip-making function is presented as a power function in the form

(47)

$$f(kT_{ij}) - bkT_{ij}^{b} \ .$$

By the substitution,

(48)

$$\frac{df(kT_{ij})}{df\,T_{ij}} = bkT_{ij}^{b-1} \ ,$$

the utility-maximizing number of trips from origin (i) to destination (j) by individual (k) can be shown as:

(49)

$$kT_{ij}^{\cdot} = \left(\frac{kM_j}{r}\right)\left(\frac{P_j^{[1/(1-b)]}}{\sum_{j=1}^{n} \frac{P_j^{[1/(1-b)]}}{d_{ij}^{[1/(1-b)]}}}\right)\left(\frac{1}{d_{ij}^{[1/(1-b)]}}\right) \ ,$$

from which the following conclusions can be drawn. The number of trips (k) from (i) to (j) will be:

1. directly proportional to the total distance traveled by (k) to all destinations

2. directly proportional to the population of each destination raised to the power $[1/(1-b)]$

3. inversely proportional to the constant

(50)

$$\sum_{j=1}^{n} = \frac{p_j^{[1/(1-b)]}}{d_{ij}^{[1/(1-b)]}} \ ,$$

4. inversely proportional to the distance between the origin (i) and (j) raised to the power $[1/(1-b)]$.

Niedercorn and Bechdolt related their findings to the gravity model in the following manner. The traditional gravity model retains the form,

(51)

$$T_{ij} = \frac{\alpha P_i^{\beta} P_j^{\gamma}}{d_{ij}^{\delta}} \ ,$$

which can be reconciled if the assumption that (M_i) is proportional to the population of the origin (P_i) raised to the power (p) is made.

Niedercorn and Bechdolt observed that for middle-class groups, time is generally the critical constraint for intraurban travel, but money is the critical constraint for travel to areas outside the region. With regard to the poorer classes, money is the major constraint for both types of travel. Niedercorn and Bechdolt believed that constraints must be determined for each area of study chosen in order to predict more accurately the spatial patterns of consumer behavior.

The effectiveness of Niedercorn and Bechdolt's study was weakened by the absence of any theoretical explanation of the trip utility function $f(kT_{ij})$. Empirically determining the values of (k) and (b) offers no explanation as to why the function takes the form $\ln(kT_{ij})$ or $(kT_{ij}{}^{-b})$. Their major contribution to the gravity concept, however, was the application of monetary and temporal constraints upon the model and the illustration of how

> . . . the so-called "gravity law" of spatial interaction can be logically derived from the economic principle of utility maximization, rather than from the vague and irrelevant concepts of social physics. The number of trips taken from a given origin to a particular destination per unit time is the sum of the number of trips per unit time that maximizes the utilities of spatial interactions of the individuals at the origin, subject to some relevant constraint. (Niedercorn and Bechdolt, 1969, p. 281)

Emphasis upon Interregional Flows

This entire period was characterized by the introduction of various gravity models into economic models (Theil, 1967; Leontif and Strout, 1963), with particular emphasis upon interregional flows. Artle (1961) had previously shown that the use of the economic base theory was inadequate for the analysis of regional economics but was able to utilize an input-output model. Nevertheless, the major difficulty encountered by previous investigations was that the gravity models had been developed only for individual commodity flows. In order to find application in regional planning, these flows must be interrelated. Wilson noted the failure when he stated:

> . . . This whole area of work falls between geography and economics: the development of the appropriate concepts of spatial interaction stems mainly from geography (helped by contributions from social physics), and the development of the input-output framework stems . . . from economics. It seems necessary to develop an integrated model. (Wilson, 1970, p. 197)

The gravity model in the form of a production-constrained model was used to study shopping patterns by Lakshmanan and Hansen (1965). Huff (1964) utilized a constrained model in predicting the probability that a resident of (i) would shop in zone (j). Wilson observed that

> . . . the models were used in these various ways in the early

> to mid-1960s without it really being recognized explicitly that they were all different members of the same family, and that their position in the family was determined by what might be called constraint information. It proves much easier to see the structure of this family if the models are viewed within an entropy-maximizing framework, . . . (Wilson, 1971, p. 5)

This observation unfortunately also applied to the group of studies conducted during the late 1960s, when viable trip distribution models were sought in transportation studies. Murchland (1966) illustrated the derivation of the gravity model from a maximization problem. Schneider (1967) attempted to maximize a function in the log W form. Analogies based on statistical mechanics were pursued by Sprukland (1966), Tomlin (1967), Loubal (1968), and Sasaki (1968).

V. REFORMULATION AND THE RETURN TO PHYSICS

From the midst of these empiricists appeared A. G. Wilson. His basic contention was that it was too difficult to work at the microscopic or individual level, even if it were possible to specify many of the individual properties of a large number of people. Wilson believed the only method which could handle urban or regional systems was one based on a macro-level where an understanding of the whole human spatial system is sought.

Statistical Mechanics and A. G. Wilson

Wilson turned to physics in general and statistical mechanics in particular to analyze urban systems. Aware of the previous sojourns by social scientists into the realms of physics, Wilson avoided spurious analogies. He believed that "when concepts are transferred from physics to social physics, the underlying analogy is often unsound, but by systematic analysis, the analogy can be made sound and the increased understanding thus achieved can create powerful tools and concepts" (Wilson, 1968b, p. 159).

Statistical mechanics provided Wilson with the inspiration for a conceptual reformulation of the gravity concept of human interaction he sought. The entropy concept provided the means. Wilson applied this concept to the analysis of human interaction because it enabled the model builder to handle especially complex systems in an internally logical manner.

Concept of Entropy

Entropy, "the most likely state," formed the heart of Wilson's work. Wilson viewed entropy as having four major characteristics: (1) it is related to probability and uncertainty; (2) it is statistic of probability distributions; (3) it is related to Bayesian inference; and (4) it has a measurable system property. Perhaps entropy can be best understood, however, as a statistical averaging method.

Gould (1972) provided an excellent explanation of entropy: Assume that there are twenty infinitesimal particles within a system which can assume various levels of activity or excitation depending upon how much energy is available to the system. The levels of excitation that the particles are able to reach are a function of the available energy. The total amount of energy acts as a constraint, so that all levels of activity are not possible. Similarly, the configurations that all twenty particles can assume together will also be constrained by the energy available.

If the energy level is assumed to be forty-two units, then the particles must be using the entire amount of the energy. If the configurations of the particles are aligned in ascending order, the number of all possible combinations within the limits of the energy value is equal to

(53)

$$W_a \frac{n!}{\pi_i n_i!} \quad ;$$

for our example, this equals 775,975,200.

If the particles are grouped according to a descending sequence, the number of all possible combinations is equal to:

(54)

$$W_d = \frac{20!}{6!4!3!2!2!1!1!1!} \quad \text{or} \quad 5{,}866{,}372{,}512{,}000 \ .$$

Thus the descending ordering has a much greater probability of occurring at any given time period. Therefore, for any reasonably *large system,* the most likely state, or those practically identical with it, will form a descending series in which the distribution of particles in various levels of activity can be described by:

(55)

$$p_j = p_O e^{-b_j} \ .$$

This expression states that the particles in an excitation level (j) will be a simple negative exponential function of the particles in the ground state, (p_O); the excitation level (e_j); and some parameter (b). Basically, this concept represents Boltzmann's Law, which describes, among other things, the most likely state, radioactive decay, light absorption, and numerous configurations in the social and behavioral sciences. The parameter (b) is a simple function of the average energy. As the energy level increases, (b) decreases in value. Defining entropy therefore involved maximizing the equation to find the most likely state.

Despite the fact that

> . . . the study of spatial interaction has been important for a long time in the social sciences, especially in geography but also in economics and sociology . . . models have been built

> which relate to a wide range of spatial interaction phenomena, among the models used, the gravity model has perhaps been used most of all (Wilson, 1971, p. 1),

the gravity model remained without a firm theoretical foundation. Wilson sought and found the needed conceptual framework in the theories of statistical mechanics.

Wilson began with the simplest gravity model based upon direct analogy with Newton's law of the gravitational force:

(56)

$$F_{ij} = r \frac{m_i m_j}{d_{ij}^2} \quad ;$$

(F_{ij}) between two masses (m_i) and (m_j) separated by a distance (d_{ij}), where (r) is a constant. In terms of a transport gravity model, the resultant equation is:

(57)

$$T_{ij} = k \frac{O_i D_i}{d_{ij}^2} \quad .$$

Trip interaction is proportional to (O_i) and (D_j) and inversely proportional to the source of the distance (d_{ij}). One critical factor, however, has been neglected by previous investigators. The fact that (O_i) and (D_j) must operate within some constraint was ignored by Wilson before the investigation. The constraints upon (T_{ij}) are as follows:

(58)

$$\sum_j T_{ij} = O_{ij} \qquad \text{and} \qquad \sum_i T_{ij} = D_j \quad .$$

In effect, the row and column sums of the trip matrix equal the number of trips attracted. The constraints can be satisfied if a set of constants (A_i) and (B_j), quantifying the zones of production and attraction, are added to form: $T_{ij} = A_i B_i O_i D_i (d_{ij})$, where:

(59)

$$A_i = [\Sigma\, B_j D_j f(d_{ij})]^{-1} \qquad \text{and} \qquad B_j = [\Sigma\, A_i O_i f(d_{ij})]^{-1} \quad .$$

(d_{ij}) must be a general function of distance, since there is no theoretical basis for keeping it as an inverse square when not within the context of Newtonian physics. Within this framework, (k_{ij}) is a general measure of impedance between (i) and (j) and can be expressed in terms of travel cost, time, and so forth. Thus $T_{ij} = A_i B_j O_i D_i d(d_{ij})$ describes a gravity model based upon a derivation by analogy with Newton's gravitational law (Wilson, 1967).

Reformulation of the Gravity Model

Wilson was not content with a formulation based solely

on analogy with gravitational laws. Rather, he sought a concept of human spatial interaction based first upon principles focused within a different branch of physics—statistical mechanics, which would have a firm theoretical base. Wilson's success in achieving this goal was due in part to his ability to define sets of variables which would completely specify the system under analysis and to construct all constraints operating upon these variables. Thus, using the variables (T_{ij})—the number of trips between (i) and (j)—a set of T_{ij}'s—$\{T_{ij}\}$—defines a distribution of trips. A state of the system can be defined as any one of the numerous ways in which a trip distribution occurs at a microscopic level. It is crucial to remember that it is the *distribution* of the trips that is of importance and not individual states. The probability of a distribution $\{T_{ij}\}$ occurring is proportional, however, to the number of possible states the system is able to achieve while in the process of producing the distribution $\{T_{ij}\}$. Thus if $w(T_{ij})$ is the number of possible combinations of the array of individuals in producing $\{T_{ij}\}$, then the probability of $\{T_{ij}\}$ occurring is proportional to $w(T_{ij})$ and the total number of such arrangements is equal to $\Sigma w(T_{ij})$. Using the concept of entropy, Wilson explained that there existed one distribution $\{T_{ij}\}$ for which $w(T_{ij})$ dominated all other terms relative to $\Sigma w(T_{ij})$, and thus is expressed as the most likely stated or the most probable distribution. This concept provided Wilson with the necessary probabilistic foundation for developing the gravity model based upon theories of statistical mechanics. It is interesting to note, at this point, that Wilson also utilizes a formulation previously developed by Schneider for the Chicago Area Transportation Study (1960).

Returning to the problem of providing a statistical theory for the gravity concept, Wilson utilized a single-purpose journey-to-work model. Remembering that the trip matrix (T_{ij}) must satisfy the two constraints:

(60)

$$\left(\sum_j T_{ij} = O_i\right)$$

and

(61)

$$\left(\sum_i T_{ij} = D_j\right) ,$$

a third constraint is added in the form

(62)

$$\left(\sum_i \sum_j T_{ij} c_{ij} = C\right) ,$$

where (c_{ij}) is the impedance or cost of traveling between (i) and (j), and replaces the objective measure of distance (d_{ij}). The constant (C) represents the fixed amount spent on trips in the regions. Its importance as a constraint will become readily apparent.

From previous reasoning, the probability of the distribution $\{T_{ij}\}$ occurring is proportional to the number of states of the system which result in this distribution and which satisfy the constraints. Nevertheless, if

(63)

$$T = \sum_i O_i = \sum_j D_j$$

is the total number of trips, then the number of unique arrangements of individuals which give rise to the distribution $\{T_{ij}\}$ is $w(T_{ij}) = (T!)/\pi_{kj}T_{ij}!)$. The total number of possible states is then $w = \Sigma w(T_{ij})$. The maximum value of $w(T_{ij})$ dominated all other terms to the extent that its distribution $\{T_{ij}\}$ becomes the most probable distribution. The problem now is to determine the method for obtaining this maximum. Thus, to obtain the set of (T_{ij}'s) which maximizes $w(T_{ij})$, subject to the stated constraints, the function (M) must be maximized, where:

(64)

$$M = \log w + \sum_i \lambda_i^{(1)}\left(O_i - \sum_j T_{ij}\right) + \sum_j \lambda_j^{(2)}\left(D_j - \sum_i T_{ij}\right) + \beta\left(C - \sum_i \sum_j T_{ij}c_{ij}\right) .$$

In this case, $\lambda_1{}^{(1)}$, $\lambda_2{}^{(2)}$ and β are Lagrangian multipliers, which have the effect of balancing changes within the system. Wilson uses Stirling's approximation B $\log N! = N \log N - N$ (since it was assumed the numbers would be sufficiently large) to estimate the factorial terms. The (T_{ij}'s) which maximize (M), and thus constitute the most probable distribution of trips, are the solutions of:

(65)

$$\frac{\partial M}{\partial T_{ij}} = 0$$

and

(66)

$$\sum_j T_{ij} = O_i \sum_i T_{ij} = D_j \ .$$

Applying Stirling's approximation, Wilson derived:

(67)

$$\frac{\partial \log N!}{\partial N} = \log N \ ,$$

so that:

(68)

$$\frac{\partial M}{\partial T_{ij}} = \log T_{ij} - \lambda_i^{(1)} - \lambda_j^{(2)} - \beta c_{ij} \ ,$$

which then becomes:

(69)

$$T_{ij} = \exp(-\lambda_i^{(1)} - \lambda_j^{(2)} - \beta c_{ij}) \ .$$

By substituting the two constraints, the result is:

(70)

$$\exp[-\lambda_i^{(1)} = O_i \Big/ \Big[\sum_j \exp(-\lambda_j^{(2)} - \beta c_{ij}) \Big]$$

and

(71)

$$\exp[-\lambda_j^{(2)}] = D_j \Big/ \sum_j \exp(-\lambda_i^{(1)} - \beta c_{ij}) \Big] \ .$$

But, with

(72)

$$A_i = \exp(-\lambda_i^{(1)}/O_i) \ ,$$

and

(73)

$$B_j = \exp(-\lambda_j^{(2)}/D_j) \ ,$$

you obtain

(74)

$$T_{ij} = A_i B_j O_i D_j \exp(-\beta c_{ij})$$

where

(75)

$$A_i = \Big[\sum_j D_j \cdot D_j \exp(-\beta c_{ij}) \Big]^{-1}$$

and

(76)

$$B_j = \Big[\sum_i A_i O_i \exp(-\beta c_{ij}) \Big]^{-1} \ .$$

> . . . The statistical theory is effectively saying that, given total numbers of trip origins and destinations for each zone for a homogeneous person-trip purpose category, given the costs of travelling between each zone, and given that there is some fixed total expenditure on transport in the region, then there is a most probable distribution of trips between zones, and this distribution is the same as the one normally described as the gravity model distribution. (Wilson, 1967, p. 258)

Thus Wilson has given the gravity model a firm theoretical foundation based upon statistical mechanics.

Wilson also provided an enlightening interpretation of the individual parameters in terms of this meaning within the gravity concept. Noting that in statistical mechanics, as well as in spatial systems, terms which result in the most probable distribution are physically most significant, Wilson described the expression $\exp(-\beta c_{ij})$ as the preferred distance deterrence function in which the parameter is determined in theory by the cost constraint equation for (c).

Nevertheless, it remains closely related to the average distance traveled—the parameter used by Schneider (1959). The greater the value of (B), the less is the average distance traveled. Thus as (C) increases, distances traveled increase and the value of (B) declines. Wilson explained the role of the (A_j's) as reducing all types slightly to compensate for increased movement to zone 1. Thus (A_j) can be viewed as a competition term which reduces most trips due to an increased attractiveness of one zone. The denomination of (A_i) is also commonly used as a measure of accessibility; an increase in (D1) would increase the accessibility of the opportunities at zone 1. Wilson's analysis established a competition-accessibility interpretation of the (A_j's) as well as the (B_j's) for changes in (O_i). This is the new gravity concept based on the most likely state theorem of entropy. Gould perceptively observed that Wilson's work "raises the gravity model phoenix-like from the ashes" (Gould, 1972, p. 696).

Concept of "Relaxation Time"

Gould (1972), however, believed that Wilson's greatest contribution to the concepts of human interaction resulted from Wilson's insight into the problems of systems that are not in equilibrium. Under these conditions the model of the most likely state and the actual system do not fit. The real world model may not be in its most likely state. Thus, if two such systems join, it may take some time to reach an equilibrium. The time required is referred to as its relaxation time.

The concept of relaxation time opened up the possibility of bringing an important, but hitherto ill-fitting and intractable, area of geographic inquiry under a larger conceptual umbrella. When two systems are coupled (cultural, political, economic, religious, linguistic), elements from each diffuse and mingle to form new equilibrium states, because their initial contact formed a most unlikely state. Similarly, if new information is injected at a certain point into a region, such a concentrated node of new ideas represents a most unlikely state. Slowly, over the relaxation time of the contact system, information diffuses, trying to reach the maximum entropy of the system. This spatial pattern could be the pattern of potential adopters, their mean information field, the distribution of psychological resistance, and the probabilistic barriers of classical diffusion theory—no more than constraints on innovation and information systems seeking their most likely configurations (Gould, 1972).

Wilson's reformation of the gravity concept provided a conceptual framework for urban and regional planners. It allowed for the study of the behavior of macro-variables in large complex systems, as a preliminary to making micro-models dynamic in character. The entropy-maximizing concept is essentially a model-building approach associated with a body of mathematics which has already been applied to dynamic systems within physics.

VI. FUTURE DIRECTIONS OF THE GRAVITY MODEL OF HUMAN INTERACTION

Now that the gravity concept of human interaction has been given a theoretical base in the probabilistic atmosphere of statistical mechanics, what of the future? If the past has been any indication, one must look toward current advances being made in physics. One visionary, Walter Isard, has suggested that concepts from general relativity physics be examined. He insists that

> . . . the different rates of change (growth and decay of relevant magnitudes) and *change* in these rates (acceleration and deceleration) to which these cultures are subject, and their different paces (velocities) and accelerations lead them to have different perceptions of space and time, to measure phenomena quite differently, and to attach different weights to variables at play in decision situations. I therefore insist that a good multiregion conceptual framework must avoid the assumption of unaccelerated regions. (Isard, 1971, p. 17)

All location and regional theory, according to this argument, needs to be revised so that any mass can be conceived of as a force field which allows for the forces of agglomeration. Variants of the gravity model are inapplicable to the reality of accelerating societies (Isard, 1971).

Isard proposes that

> . . . we leave the world of scalars, vectors and move into the world of scalars, vectors, and tensors. We must leave the world of Euclidean space and enter the world of Riemannian space—a space which is suited for the curvilinear coordinate systems of accelerating relative to each other—a world of which relationships expressed among tensors are preserved with transformation from one curvilinear system to another. (Isard, 1972, p. 18)

Wilson might reply, however: "Analogy may generate insight, but it does not contribute to validation" (1969).

VII. CONCLUSION: THE NATURE OF PROGRESS IN GEOGRAPHY

The close relationship between the evolution of theory in physics and the evolution of the gravity concept of human interaction is apparent. As in any discipline whose boundaries are ill-defined, peripheral influences are frequent. In the case of the gravity concept, an interesting pattern emerges.

The gravity model's conceptual basis came into being after the development of Newtonian physics. Early theoretical postulates were developed by astronomers and physicists. After a time lag, advances were made within the geographic community in the form of a series of empirical studies. With advances being made in the social sciences, in particular economics and behavioral psychology, applications of the gravity concept were reformulated. Again, however, the early research originated outside of geography. This can be explained in part by the fact that during this period geographers were engulfed in the controversy over uniqueness and the imagined difficulties of a dualistic system of data analysis. While related disciplines were utilizing spatial frameworks to explain related phenomena, geography was not. Even the planning sciences (especially the transportation and urban studies groups) made substantial contributions to the concept of human interaction.

The Nature of Progress in Geography

In general, geographers' contributions to the theoretical evolution of the gravity concept were minor. These contributions tended to be nothing more than testable hypotheses after the basic formulations had been presented. This hesitancy on the part of geographers and neogeographers to explore other avenues of research is only interrupted on occasion, when such men as Isard and Wilson are able to inject themselves into the mainstream of research being done in the physical and the behavioral sciences. It is not so much a question of borrowing ideas from related fields to be utilized in geography as it is a simple awareness of the progress of science and scientific methods of analysis. The entire question of the role that geography plays with an evolving philosophy of science has been ignored too long. If geographers are to contribute substantially to the knowledge of the complexities of human spatial interaction, then they must be prepared to become part of the community seeking basic principles, as are researchers in physics, chemistry, and biology. With an increase in the rate of scientific progress as well as an increase in the complexity of the research, it is all the more necessary to be part of that community if the relative time lag is not to condemn progress in geography to an antiquarian hand-me-down status.

The evolution of the gravity concept was a complex process, but a clear pattern of evolution is visible. In its most basic form, it consisted of a progression of advances based upon a time lag after the basic discoveries were made in the physical or behavioral sciences. Each advance was followed by a series of empirical studies, which resulted in basic reformulations setting the stage for the next adoption. Most important of all is the fact that this process is still operating today, under conditions of accelerating change that carry with them the danger of running faster and falling further behind.

5

IMPLICATIONS OF "SCHOOLING" IN ECONOMIC ANTHROPOLOGY FOR INTERPRETATIONS OF THE ECONOMIC GEOGRAPHY OF NONINDUSTRIAL SOCIETIES

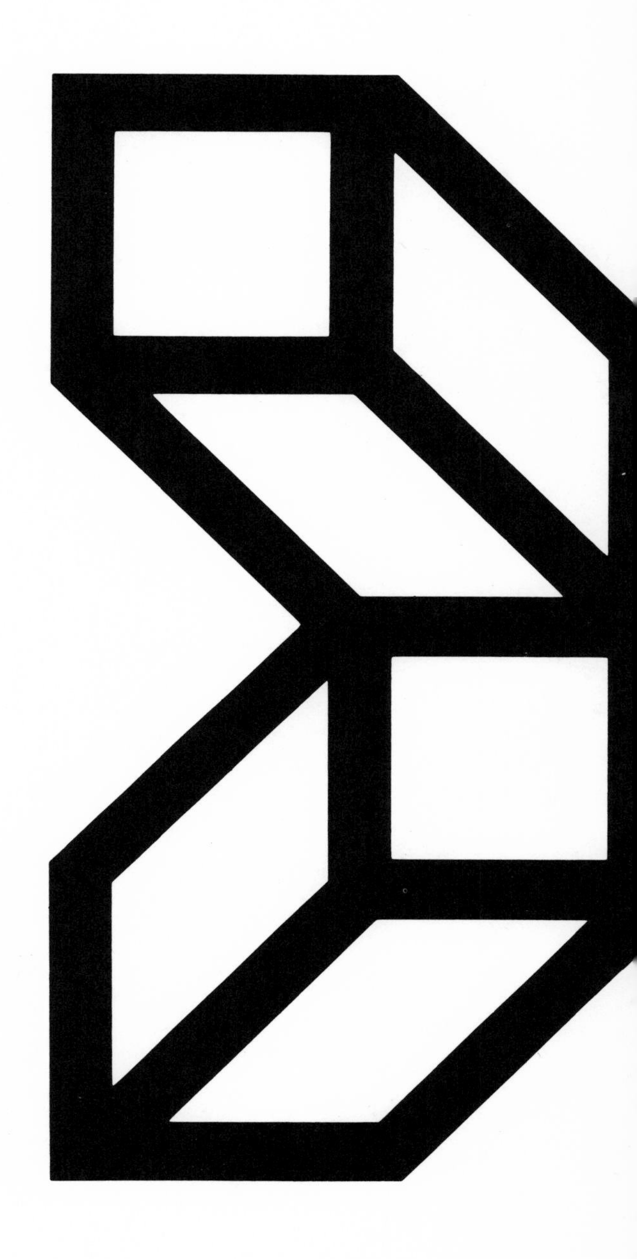

Donald Jones
University of Chicago

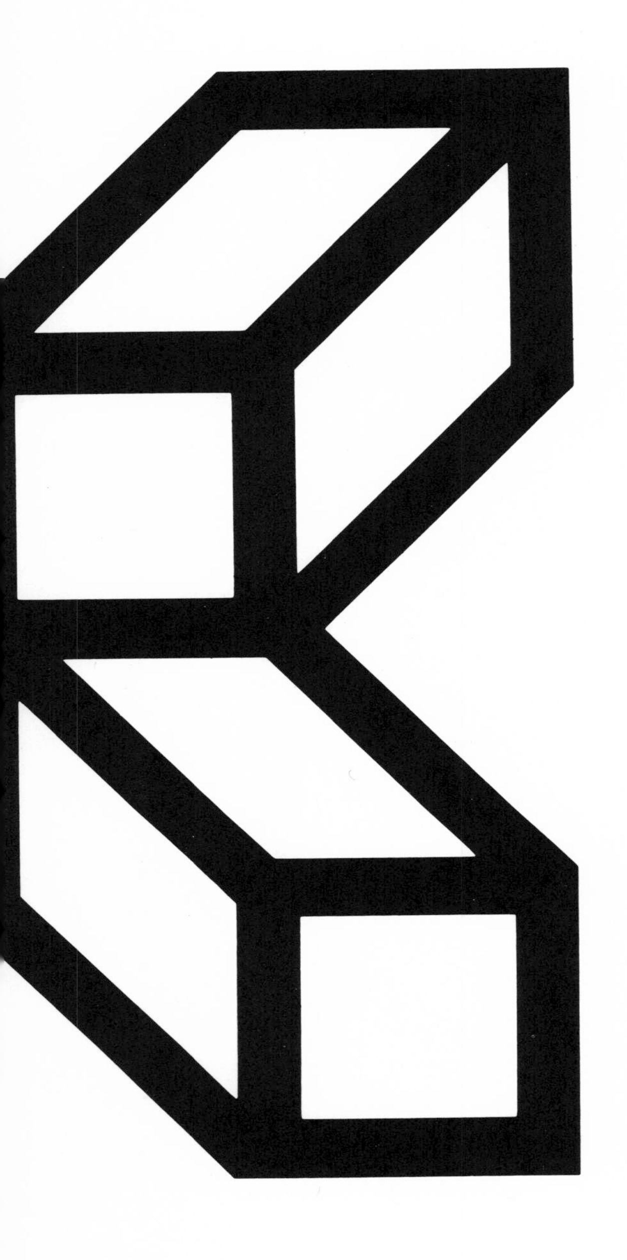

IMPLICATIONS OF "SCHOOLING" IN ECONOMIC ANTHROPOLOGY FOR INTERPRETATIONS OF THE ECONOMIC GEOGRAPHY OF NON-INDUSTRIAL SOCIETIES

The hand-me-down pattern highlighted in the preceding essay reappears in other parts of the discipline. Cultural geographers have devoted the bulk of their attention to peasant and/or non-Western, nonindustrialized societies, often using concepts borrowed from anthropology. Their treatments of matters economic among these peoples have, accordingly, relied heavily on intellectual traditions of that discipline, and the economic geography of nonindustrial societies is heavily influenced by the particular strand of thought in economic anthropology known as substantivist. The analytical quality of the economic geography grounded in this approach to economic life is not particularly penetrating. Wagner (1960), for example, relies on Polanyi's typology of economic transactions, and that scholar's peculiar philosophy of the economy, to describe the economies of the particular societies with which he (Wagner) is concerned. Wagner's discussion of ways of livelihood is another descriptive classification in which the analysis of economic activity remains, essentially, at the level of a day in the life of a predator. Such phenomena as the interrelationship of a particular level of technology, or of a certain societal attitude toward property rights, and the group's selection of production/cultivation practices are not addressed.

In the present paper, I propose to examine the approach to economic anthropology upon which the economic geography of nonmodern societies has been based. Following this analysis, I will take up one particular topic for which that attitude toward the analysis of economic behavior has produced no satisfactory solution, using a different method of analysis to produce what seem to be somewhat more satisfying (or at least promising) results. Both the discussion of the underpinnings of the contemporary economic geography of nonmoderns and the analysis of primitive exchange suggest the fruitfulness of some paradigm shift in that corner of cultural geography.

I. TWO SCHOOLS OF ECONOMIC ANTHROPOLOGISTS

Current thought in economic anthropology can be viewed as comprising two broad schools who divide themselves on an axis revolving around attitudes toward the use of Western economic theory in the study of nonmodern (or even modern, non-Western) societies. Those scholars who have pursued, to some extent, the application of economic theory in their studies have become known in the last two decades as formalists. The range of acceptance of economic theory by writers who have been so branded is so wide that the validity of their common grouping rests on their common behavioral trait of not rejecting that theory out of hand.

There is probably a stronger feeling of solidarity among economic anthropologists of the substantivist persuasion. Ideological undercurrents run through this literature, reducing the effectiveness of communication between substantivists and formalists. The mainsprings of the substantivist intellectual position derive from Karl Polanyi's perceptions of economic theory (1944), which are therefore the central object of the present examination. Polanyi's ideas have been received with few or no reservations by the students of the substantivist tradition, but the more imaginative of these current writers, while paying lip service to Polanyi, are, if their rhetoric is ignored, pursuing ideas with quite different implications. After the discussion of Polanyi's original message, some theoretical work of Marshall Sahlins is examined. The central propositions (or logical implications, which are not clearly extracted by that author) of Sahlin's theory of primitive exchange are joined together and formalized in a unified economic approach.

Substantivism

George Dalton's 1969 statement regarding the usefulness of economic theory to anthropology is the most precise summary of the substantivist position regarding the appropriateness of using economics in anthropology.

> Some anthropologists seem not to understand that conventional economics—the most abstract and mathematical of the social sciences—does not deal with what anthropologists mean by human behavior, and that the concepts of conventional economics relating to economic organization are not (with some minor qualifications) fruitfully applicable outside of market systems.

The tone of the statement is a reasonably accurate reflection of the dialogue between substantivists and formalists on this issue. Particularly significant is the dual appearance of *anthropologists* in the sentence, qualified on the first occasion by the adjective *some,* suggesting that *some* in that instance is contrasted with the implicit modifier, *all,* in the second sentence.

A significant part of the communication problem between substantivists and formalists (and, in general, between economic anthropologists and economists) turns around the contrast between the primarily empirical approach of anthropology and the logico-deductive methods of economic theory. The anthropologist is apt to want empirical demonstrations of the more rarified assumptions of economics and has a tendency to try to counter an entire logical argument with an empirical disproof, a not entirely satisfactory methodology. Such confusion of logical (or at least definitional) issues with empirical issues will be pointed up in the discussion regarding scarcity.

It seems to this writer that the most immediate intellectual touchstone in the substantivist position is the work of Karl Polanyi on exchange types. In Polanyi's work arise concerns over the legitimacy of transferring to non-Western, nonmodern (nonmarket) economies the notions that scarcity exists and hence choices must be made, and that prices and, hence, markets, exist in these societies. Intimately involved with these issues is the limit of the economy—the extent to which an economy can be identified separately from other social organs of the society. We will examine these issues in order of their cruciality to one another and to the substantivist position as it is presented below.

Exchange Types and the Limits of the Economy

The distinguishing tenet of the substantivist position, that Western economic theory is not applicable to nonmodern societies, receives its authority from Polanyi's distinction of three exchange types (reciprocity, redistribution, and exchange), and his assignment (relegation) of the applicability of economic theory only to the last type. From a series of definitions, some only implicit, regarding the exchange type, all the substantivist positions follow. A major difficulty is that these definitions often involve previously defined terms, and consequently the argument degenerates into castigation of the newly defined straw men. These exchange types, in turn, depend upon Polanyi's definition of the economy, which definition itself rests upon what that writer calls the formal/substantive distinction. Finally, that distinction is based upon a further division between material and nonmaterial wants. To begin demolition of the use of the exchange types as a basis for rejecting the use of economic theory, it will be best to proceed by showing the importance of the material/nonmaterial distinction between kinds of wants, then demonstrating that this distinction is particularly misleading.

Though exchange types are said to be not mutually exclusive—that is, more than one type may exist within a society at the same time, though presumably a single transaction must fall within the realm of only a single type—their separability depends on dichotomies derived from the limits of the economy argument in a very special way. That is, if the definition of the economy excludes from one of the exchange types 90 percent of transactions which occur in the society, and if, by that definition, certain forces (processes) come into play only in the presence of the excluded exchange type, then one has well-nigh eliminated that force as an important (even relevant) factor in that society.

To Polanyi, the formal meaning of the economy is the set of economizing (maximizing) rules, based only in logic and having no necessary connection with fact. It derives from or requires, (1) insufficiency of means, and (2) alternative ends—two conditions which Polanyi holds are empirical [conventional or technological (Polanyi, 1957, p. 246)] rather than logical or definitional. These optimizing rules, when applied in an economy of the third type, the market exchange economy, become economic analysis; so, by definition, Polanyi has precluded the applicability of economic theory to primitive or archaic societies, in which the market is absent. (Polanyi's criteria for evaluating the presence or absence of the market, though relevant to the argument at this point, particularly inasmuch as these criteria reinforce absence of the market from nonmodern societies [load the dice, so to speak, against using economic theory in such cases], will be examined below.) Though this proof by definition is, in itself, sufficient ammunition for the contemporary substantivist position as it has been developed, the case against the anthropological use of economic theory is rather messily reinforced by ideas coming out of the definition of the substantive economy.

In contrast to the formal meaning of the economy, the substantive economy is the actual going about of everyday business—the acquiring of the necessities of life. According to Polanyi:

> The substantive meaning implies neither choice nor insufficiency of means; man's livelihood may or may not involve the necessity of choice and if choice there be, it need not be induced by the limiting effect of a "scarcity" of the means; indeed some of the most important physical ["material," this writer presumes] and social conditions of livelihood such as the availability of air and water or a loving mother's devotion to her infant are not, as a rule, so limiting. The cogency that is in play in the one case [the "formal" meaning of the economy] and in the other [substantive] differs as the power of syllogism differs from the force of gravitation. The laws of the one are those of the mind; the other are those of nature. (1957, pp. 243–44)

It is important to suggest, in reference to the last sentence of this passage, that the force of gravitation is as much a law of the mind as is the syllogism. It appears that the existence of different types of abstraction persuades Polanyi that one of them must be the real thing. His subsequent choice of a dividing line between that real thing and the figment produces a pair of infelicitous distinctions. The first of these, the formal/substantive dichotomy, fosters the second, the material/nonmaterial division of wants, this latter distinction permitting Polanyi to consider scarcity an empirical issue. Specific issues which this passage raises are: (1) What can produce the necessity of choice? (2) To what extent is the existence of scarcity an empirical issue, and to what extent a definitional matter? (3) What are the implications of the material/nonmaterial distinction implicit in the "air and water"/"loving mother's devotion" example of needs? Fuller exposition of these issues will put flesh on the bones of the argument as it now stands.

Valuation in the Exchange Type

Before proceeding, we should give definition to these exchange types to which we have been referring. We can then work through some analyses of a particular transaction using these exchange types as a framework, in order to test the analytical utility of Polanyi's construct.

The first exchange type, reciprocity, involves movement of goods or services between "correlative points of symmetrical groupings" (1957, p. 250). Redistribution requires movements into, and out of again, some center. And finally, "exchange refers here to vice-versa movements taking place as between 'hands' under a market system" (p. 250). The vagueness of this definition is easily seen; in fact, the reader's intuition of what a market system is seems to be Polanyi's principal reliance for definition, though he does clarify somewhat by noting that the market is characterized by disharmony and antagonism, while the other exchange types generally are associated with goodwill and public spirit.

On page 251, we are told that exchange only produces prices if such transactions occur "under a system of price-making markets . . ." Since price formation contains a valuation process, the absence of prices in the market types of reciprocity and redistribution leaves these latter two without this valuation process and does not replace it with any other, unless one wishes to offer Nash's concept of "near equivalence" (Nash, 1966, p. 31) which means that, "What I give you is close enough (in value-personal worth or estimation) to what you give me that we won't worry about the difference, because if we trade enough times the difference will probably wash out." Pursuing such a valuation concept, assuming that such is implicit in Polanyi's writing, there is no necessary difference between reciprocity and redistribution other than a centrality of some one unit, which, of course, depends upon one's point of view.

Let us consider some peasants and a king. Suppose that each peasant gives the king a pig or a chicken in order to keep from getting his head lopped off (by local marauders from whom the king protects the peasants in return for tribute, or by the king if the peasant is caught not rendering). Why should this swap be either reciprocal or redistributive? Each peasant has a reciprocal relation with the king—protection for tribute. Alternatively, there is a redistributive relationship, if one does not include the protection in his accounting scheme—tribute for nothing; the king can give the tribute to his court, another king, or other peasants, or simply consume it himself. If simple accounting differences will produce different exchange types from the same act, these exchange types do not discriminate well. (This is different from Polanyi's claim that the types are not mutually exclusive.)

Let us consider a valuation process in this transaction, supposing that the form of tribute is animals. Each peasant will roughly estimate his chances of getting his head lopped off by a marauder in the absence of the king's protection and will assign a (subjective) value to that probability; that is, there is a certain amount of goods he would be willing to surrender to insure himself against the risk of untimely death. If he wishes to avoid the possibility of being beheaded at all costs, he will certainly give that one pig or chicken—and consider that he received a bargain. If his computation tells him, however, that his chances of getting killed by a marauder are slim, and if he is willing to accept some (culturally and personally determined) amount of risk, he may decide that the value to him of the chicken is greater than the value of the king's protection; in this case he would keep his chicken. Now if he had to consider the finite probability that the king would kill him for that chicken, he must add this probability to the probability of being killed by the marauders and subtract the sum from one to determine his probability of staying alive; he must now compare his valuation of the chicken against his valuation of insuring against this greater risk of life termination.

From the king's point of view, he must decide how many chickens to extract from his peasants. Balanced against this wealth redistribution from peasant to king is the increasing chance that the peasants will get fed up and revolt, or that they will not have enough to eat and will consequently be able to supply him with fewer chickens and pigs at later dates. (All wise rulers keep an eye out to the first possibility, only if insomuch as to insure themselves against that risk by foregoing some of the redistributed wealth to maintain a personal guard and/or an army.) Taking into account the peasants' relative valuations of the pigs, chickens, and their own lives, as well as the effectiveness of the marauders in reducing the peasant population, the king will figure out that at some level of pig and chicken taxation he will receive a lower net tax revenue, due to tax evasions and collection costs, if he increases the rate a bit, and a lower net revenue, due simply to the rate, if he lowers the rate a bit. If both king and peasants are smart as well as wise, they will find themselves with that tax rate of chickens and pigs which maximizes the king's tax revenue, subject to the peasant's preferences. It so happens that this tax rate will also be the one which makes the peasants happiest, since they have the amount of antimarauder insurance they wish to hold. Is this reciprocity, redistribution, or exchange? If this question could be answered unambiguously (or even ambiguously), would the knowledge of the proper classification be worth having?

Now consider a pair of peasants, one with only chickens and the other with only pigs. If the one with only chickens would like to have a pig, and vice versa, they may eventually find an acceptable ratio of chickens per pig. If they find such a ratio, it will look like this: pigs per chicken for the man with chickens, times the ratio of the man's subjective valuation of a chicken to his subjective valuation of a pig, must equal the same calculation for the man with the pigs; and the chicken per pig ratio must be the same for both men. The processes of valuation and change of hands is known in economics as exchange. From Polanyi's definitions, one cannot distinguish this from reciprocity, though such a verdict would implicitly give Polanyi the benefit of the doubt on the inclusion of a valuation process, which is the only method by which culture may be introduced into such a transaction. But in disallowing this benefit of the doubt, Polanyi leaves the impression that different *forces* are at work in his exchange types, in which case it is not clear that the differences in the *forms* which the types occupy are of importance. Were we to make one of our peasants a king, all three exchange types could fit the same transaction.

Going back to the two peasants trading chickens and pigs, suppose we wished to develop the case in which the two were good friends, and the values of the traded items were not exactly equal. Does this imply that the valuation process is therefore qualitatively different from the one described above? It seems plausible to suggest that the "losing" individual's pattern of indifference between chickens and pigs is widened into a band rather than a single line, or possibly that some goodwill is being exchanged. In the second case, we must know the individual's ordinary chicken/pig valuation before we can assign some portion of the final exchange outcome to the objective resource cost of, say, the pig in terms of chickens, with the residual going to goodwill. At any rate, the amount of goodwill that can be given away has physical limits (that is, a person cannot give away what he does not have), and it seems quite likely that a person will offer only so much of the fruit of his labor for goodwill; he must eat something, although good feelings may certainly alter his appetite. The ordinary exchange framework of economic theory has the advantage of being capable of analyzing the effects of the finiteness of the world, while Polanyi's three-category, exchange type classification appears to offer no analytical compensation for the surrender of that advantage.

LOVE AND THE ALLOCATION OF TIME

To a Polanyi disciple, the "human economy" is

> an instituted process of interaction between man and his environment, which results in a continuous supply of want satisfying material means. Want satisfaction is 'material,' if it involves the use of material means to satisfy ends; . . . (Polanyi, 1957, p. 248)

It is this material limitation that permits Polanyi to make scarcity an empirical matter.

Taken separately, individual items may be available in such large quantities that everyone's desire for possession or use of it may be satiated without exhausting the supply. In such a case, a zero price may be assigned to the item. Let us consider some primitive group, for whom every item desired is described by such conditions. Let us assume that some time period exists over which to measure these desires; for example, the people want food every so many hours. It is during this time period that the satiation of desires without exhaustion of supplies must occur to fulfill Polanyi's nonscarcity condition. It is conceivable that it may take longer than this appropriately defined time period to collect the amounts of each of these items required for total satiation. A choice must be made regarding how much of what free goodies to collect in order to obtain the greatest satisfaction possible, short of satiation. This may still be considered a candidate for empirical verification, however, but the interaction between nonmaterial time and material goods consumption does point us in a direction which suggests the usefulness of merging the distinction between material and nonmaterial means.

The citation above describing "a loving mother's devotion to her infant" as a "social condition of livelihood" establishes the former as a "nonmaterial means," which will not ordinarily involve scarcity. This immediately takes us into a morass of definitions. What is an effective definition of love? "Effective," in this case, implies that the definition we choose highlights the aspects of the term that are most pertinent to answering our question. How would we measure love, and how would we define and measure the particular "condition (or conditions) of livelihood" which love helps to attain? Are the satisfactions of these conditions all-or-nothing satisfactions? Can one be more or less satisfied with the outcome of the application of a particular amount of love toward some "social condition of livelihood"? These are questions which Polanyi's writings raise, but they are squarely faced by neither Polanyi nor his disciples. These issues simply cannot be addressed with the Polanyi framework, so let us push on with the economic framework, which Polanyi tells us is inapplicable, and find out what it tells us.

The mother herself will experience a range of wants, some of which may be altruistically motivated and have little to do with physical materials, others of which may be more mundane. The wants the mother faces may be sufficiently numerous that they impinge on the simultaneous satiation of all of them. ("Satiation" is used rather than "fulfillment," since Polanyi does not suggest gradations, or degrees, of the fulfillment of wants.) There is no reason to suppose that the utilization of the purely nonmaterial means for one purpose will not conflict with the application of the same nonmaterial means to some other end; that is, there will not be enough nonmaterial means to go around. This mother's love may not be sufficient to satiate several children and a mate. When one further considers the acquisition (production) of those means, material or otherwise, one must ask whether there is substitutability among them. The effect called love is produced by (or, at least, expressed in) a number of elements, one of which is time. The mother may choose to devote a lesser amount of her resources to the furnishings of love to her children than is necessary to satiate her desire to do so.

This may seem like a logical impossibility, but let us take a closer look. The woman has a list of wants arrayed before her, the fulfillment of which is accomplished on an all-or-none basis, since Polanyi does not imply that they may (or should) be less than totally satisfied. The application of a nonmaterial mean to one purpose (end) may preclude its use in another purpose. To be able to satiate all the desires (ends), she must be able to produce an arbitrarily large number of nonmaterial means. Two possibilities may arise to preclude this. First, the provision of these means may involve physical means; for example, the woman has only so much time and energy. Second, exclusion of particular ends in the application of the means may be required to attain some other, symbolic end, so that the only way one end may be achieved is that others be foregone. This latter contains the essence of choice, which brings us to an examination of Polanyi's concept of that action.

Although he states that he has defined choice, examination of his effort discloses that he has, instead, defined "rational action," in terms of choice. [Polanyi (1957), p. 246: "having thus defined choice . . ." This passage apparently refers to the statement: "Rational action is here defined as choice of means in relation to ends" (p. 245). His statement (p. 247) that: "Assuming that choice is induced by an insufficiency of the means . . . " does not define *choice,* either; nor do the other requirements for choice: "Formal economics refers . . . to a situation of choice that arises out of an insufficiency of means . . . For the insufficiency of means to induce choice there must be given more than one use to the means . . ." (p. 246). The closest Polanyi comes to defining choice is the passage: "The formal meaning of economic derives from the logical character of the means-ends relationship. . . . It refers to a definite situation of choice, namely that between the different uses of means induced by an insufficiency of those means" (p. 243). The grammar is awkward, and cursory reading may leave the impression that *choice* has been defined as being somehow "the different uses of means induced by an insufficiency of those means." On closer reading, however, this is merely one possible type of choice which conceivably could arise.] Without specifying a firm definition of choice before discussing the conditions surrounding choice, Polanyi permits himself to wander into confusion regarding the relationships between choice, insufficiency, and economic analysis. Let us define choice to be the ability to select discrete events. To exercise this ability is to choose, or to select. The important issue regarding choice and economics is whether the events are exclusive of one another, or under what conditions and to what extent the selection of one event precludes the selection of another event. For example, one may have two options for the use of time: eating and sleeping. If one chooses the event "to eat," the event "to sleep" must be foregone, but only to a certain extent. Choice is necessarily made at that margin of time which divides sleeping and eating, simply because the world events are defined so that the individual must do one or the other. The decision to do one or the other is produced by the individual's valuation of those two events—that is, "what he likes." If, after seventeen hours of sleeping, he prefers one more hour of eating to one more hour spent sleeping, this valuation results in the choice of the event called "eating" for one hour. If after that hour spent eating, he still prefers one hour of eating to one hour of sleeping, he will make the same choice. Of course, one may define the bundles so that he can choose, at the beginning of the day, combinations of X hours of eating and Y hours of sleeping to M hours of eating and N hours of sleeping, without changing the nature of choice so defined. With bundles defined so that they are a continuous array of alternative combinations of hours spent in eating and sleeping, he will eventually reach some bundle to which he is indifferent whether he has a little more sleep and a little less meal time, or a little less sleep and a bit more meal time. At this combination of hours spent in sleeping and in eating, the increase in satisfaction he gets from a little more of one activity must be exactly equal to the decrease in satisfaction he would incur by giving up the same amount of the other activity—or close enough so that the individual making the choice would use up more satisfaction in further measurement than he expects to get from any further alterations in the eating-sleeping ratio. Why "the same amount" of that other activity? The rate at which time spent sleeping may be physically transformed into time spent eating is unity: an hour is an hour. The hour spent in sleeping may be physically transformed into no more and no less than one hour spent in eating. Hence, when the individual finally settles upon a ratio of sleeping time to eating time, the ratio of satisfaction obtained from either must be unity. So the ratio of the time spent in these two activities implicitly defines, in terms of each activity, this individual's valuation of these two activities.

Suppose, however, that this individual is satiated with both sleeping and eating before the sum of hours spent in each reaches twenty-four. The person need give up no sleep to eat, nor any meal time to sleep. There is no rate of transformation of time spent in one activity into time spent in the other. There is no cost of either activity, since the cost of, say, time spent in eating can only be measured in terms of what must be given up to obtain it, time spent in sleeping. Since we have defined only two events, and this individual must do one or the other, this description of non-exhaustion is nonsense.

If we consider time to be a valuable resource, the value of which may be expressed in terms of the most desired activity in which one can engage during a unit of time, there is a good case for considering scarcity to exist by definition without being cited for heroism of assumption. In particular, if certain sets of events (activities) may be available for choosing only during a certain period in the day, the cost of choosing a certain amount of one activity in terms of the amount of the other activity which must be foregone may be clearly seen.

What does all this say about economics and the loving mother? The allocation of her time and effort among activities depends crucially upon her own values, but the sum total of her activities is unlikely to satiate her ambitions. Her attentions to her child may inflict costs on other social elements in her society—tribal cohesion, town spirit, or vitality of the ladies' backgammon club—which may or may not be offset at a later date by the (better-socialized) child. These are economic aspects of social actions, and unless one considers a properly economic act to be antisocial, this class of activities will include the economic system.

Thus far, in the discussion of the exchange types, we have: (1) pointed to the absence of a valuation process in Polanyi's construct; (2) shown that the valuation process contained in the economic theory of exchange can replicate, with one model, transactions which are said to represent each of Polanyi's three market models; (3) given a definition to *choice,* and defined social events so that scarcity exists definitionally and without great strain on common sense; and (4) using these notions of choice and scarcity, indicated the economic aspect of actions which are not ordinarily considered economic. Let us now turn to phenomena more usually associated with matters economic: markets and prices.

Markets and Prices

Thus far the discussion has been on a conceptual level. It is useful at this point to apply the suggested conceptual approach in an examination of the mechanics of economic activity in a nonmodern setting.

Equivalency Ratios and Prices

Within Polanyi's framework, price is a form of equivalency, designating the "quantitative ratios between goods of different kinds, effected through barter or higgling-haggling"; however, price is not the only form of equivalency and, indeed, is only found in the exchange mode known as market exchange. Other equivalencies, found in the reciprocal and redistributive exchange modes, are fixed, rather than fluctuating, as are prices. The equivalencies found under redistributive transactions designate the (fixed) ratios between different goods which are acceptable in exchange for one another, or which are designated for use in particular categories of transaction. As examples, imagine that only certain items may be offered if one wishes to obtain woodpecker scalps, or that only pigs may be used to pay one's taxes. Under reciprocal forms of integration, the equivalency indicates the amount of a commodity which is " 'adequate' in relation to the symmetrically placed party" (p. 269).

When "exchange" (as opposed to reciprocity or redistribution) integrates the society, prices are determined by the individual parties' bickering back and forth (which is probably an accurate interpretation of higgling-haggling) until they agree on an equivalency. Such proceedings will create fluctuations, over time, in the resulting prices. Now consider the equivalency ratio between woodpecker scalps and, say, stone ax blades; they may be exchanged only for each other. If one had a particularly ardent desire for woodpecker scalps, it would be possible to concentrate one's efforts on making stone ax blades to such an extent that one could, at the fixed equivalency ratio, obtain more woodpecker scalps than exist. What should become of the equivalency rate? Polanyi's scheme offers no suggestion about how to proceed.

So let us suppose that the exchange rate were absolutely fixed and, by societal consensus, should not be altered. Let us assume that 50 woodpecker scalps exist, that the equivalency ratio between woodpecker scalps and ax blades is one-to-two, and that our blade maker has made 150 blades. To begin with, he has no woodpecker scalps, and no one else has ax blades. He begins to obtain scalps at the one-to-two ratio. There is no societal concensus (law) requiring individuals to trade in their woodpecker scalps for ax blades, and when the blade maker has acquired 35 scalps, he can no longer convince anyone else to part with woodpecker scalps. The equivalency rate thus does not reflect the society's preferences for woodpecker scalps and ax blades. Will there develop pressures to subvert the law by secretly offering three ax blades for a woodpecker scalp? Could the blade maker lobby to change the exchange rate to, say, one-to-three or one-to-four, in order that he might entice some of the remaining 15 woodpecker scalps away from their owners? If such lobbying could occur, what would be the fate of the idea of fixed equivalencies?

If one examined actual cases of such schemes in which only certain goods might be traded for certain other goods, one quite likely would find that where the schemes worked well, the society's ability to alter the supplies of each good would be roughly the same for all the goods in that category. If such were not the case, the system would probably break down or have repercussions elsewhere in the society. Similarly, the society's preferences for the items so categorized and segregated from exchange with other commodities must remain stable. If a large portion of a society behaves as the blade maker in the preceding example, that exchange category will not long remain intact in the social organization. In a comparable manner, with reference to reciprocal transactions, any stability in equivalency ratios must depend on stability of the notion of adequacy and the relative supplies of the things exchanged.

An example of such a fixed-ratio exchange category is presented by Cancian (1965) in his study of the Zinacantan cargo system. The cargo is an "office" of religious responsibility and public service, purchased for a one-year period and yielding prestige to the purchaser. Four "levels" of cargo exist, and they must, ordinarily, be occupied in order, from least to most prestigious. Since more individuals desire the number of cargos that exist at any one time, there are waiting lists for the various offices; this queuing procedure is a partial alternative to a fluctuating (market-clearing) purchase price. Cancian portrays the cargo system as a device for redistributing wealth through the society. As the Zinacantecos have become wealthier (and more numerous), as a result of economic and legal reforms following the Mexican Revolution (1914), the queues for the individual cargoes have grown longer, and additional first-level (lowest) cargoes have been created (Cancian, 1965, p. 187). Nevertheless, pressures still threaten the existence of this institution. Longer waits for the prestige offices will have the effect of permitting participation by a much smaller proportion of the population. Cancian reports several instances of such pressures: (1) increasing participation in other social institutions that have been created as imperfect substitutes for participation in the cargo system (Cancian, 1965, p. 190); and (2) attempts to buy cargoes "out of order,"—that is, requesting to occupy second- (a higher) order cargoes before having occupied a first-order cargo (Cancian, 1965, pp. 190–91). The similarity should be noted between the structure of this latter example and that of the example constructed (above) of the ax blade maker wishing to change the woodpecker scalp-ax blade exchange rate so that he could obtain more scalps. Cancian's findings give some content to what a more skeptical reader might be tempted to call an "empty box."

The fixity of an equivalency ratio which reflects adequacy will have similar frailties. How many chickens is a king's protection worth? If one could take half as many chickens, give them to a neighbor of similar nobility to oneself, and receive the same protection that the king provided, how long would one continue to feed the king? If half of the chickens given to the king for this service represent the cost of the physical services rendered by the king, what does the other half of the chickens represent? To content oneself with saying that x number of chickens is adequate recompense for one's peer, but 2x is adequate remuneration to the king for the same delivered item is insufficent. Interesting and significant questions regarding a society's valuation of elements of their social organization are involved with the second half of these chickens, but one must be willing to separate the two payments. The pricing framework, which Polanyi rejects for such transactions, will aid one in formulating and answering such questions, without forcing any loss of information or imposing any structure or process which "does not exist." [As Polanyi himself notes, "rationality" is not at issue (though later minds have, strangely, not always heeded him on this point), and the "auctioneer" of the Walrasian *tâtonnement* process and similar heuristic constructs are simply constructs.]

Polanyi's concept of a price as a quantitative ratio between goods of different kinds is adequate, if not analytically useful, as part of his exchange type/equivalency type scheme. Nevertheless, the hypothesis that he really has a clear idea of what a price is suffers a severe setback in his discussion of types of trade (gift trade, administered trade, and market trade).

> Higgling and haggling is not part of the proceedings [of administered trade, which is not clearly separated from trade of nonmodern societies in general]; equivalencies are set once and for all. But since to meet changing circumstances adjustments cannot be avoided, higgling-haggling is practiced only on other items than price [italics are Polanyi's], such as measures, quality, or means of payment. (1957, p. 62)

In other words, the price of sugar is the same as it always was; only now you get half as much. This concept of price has its appeal, as those who witnessed the Nixon price freeze may recall in regard to packaging of fresh meat and chewing gum; however, not many people were fooled for long. Their concept of a price was the cost (in whatever unit of measure: dollars, deutschmarks, or 1958 Edsels) *per unit of quantity* of an item. And since a pound of rotten lamb chops is not readily substitutable for a pound of fresh lamb chops, the qualitative attributes of the constant quantity must be constant for the constant price tag to convey the signal that the price (price per unit, if one pleases) is the same. If one haggles over whether one is going to receive one handful of corn or two in exchange for one shekel, one is, assuredly, haggling over the price of corn.

A price is simply a ratio that is finally agreed upon for the exchange of two goods, whether it is agreed upon after bloody fighting or has remained constant with little discontent for some time. Exchange/trade creates prices, not vice versa, as Polanyi contends (1957, p. 268). ["Thus language itself makes it difficult to convey the true state of affairs, namely that 'price' is originally a rigidly fixed quantity, in the absence of which trading cannot start" (Polanyi, 1957, p. 268). The last sentence in this paragraph seems, at first reading, to be a contradiction to this statement: "Traditionally, the sequence was supposed to be the reverse: price was conceived of as the result of trade and exchange, not the reverse" (p. 268). "Traditionally" must refer to the nineteenth- and twentieth-century classical and neoclassical economists rather than to the archaic societies he has previously considered. Otherwise, he is directly contradicting himself within a space of three sentences.] Additionally, a "price" need not identically reflect "cost" to both parties of the exchange. The price of otter pelts in terms of wampum need not accurately reflect the number of otters which could be caught and skinned, per unit of time, if one chose to spend his time in that occupation. One may have found the otters already skinned and consider the cost of obtaining them to be the effort involved in stooping over to pick them up. The resulting price of otter skins in terms of wampum will be too cheap; that is, the number of otter skins which a certain amount of wampum will buy is too great, or is more than one could have collected if one spent the time he spent making wampum in otter hunting instead. In such manner, an agreed-upon price can reflect power in bargaining as well as preferences and technology; or it may reflect the cost to an individual but not the cost to his society. Thus the ratio which is observed as a "price" is an interesting social phenomenon to ponder.

This brings us to the point which is probably the one of greatest concern to Polanyi: "Who 'institutes' the 'market'?" Depending upon the social distribution of power, those equivalency ratios which we have called "prices" will reflect different forces in different degrees. If prices are arbitrarily assigned by some authority, they will reflect that authority's preferences—either his preferences for his own consumption, or his preferences for everyone else's consumption. In such a regime, however, there is no assurance that the quantity of each good demanded will equal the quantity supplied; there is the distinct chance that someone will be left "holding the bag," while someone else will be willing to pay for more than he can obtain.

In the case of kings, chickens, and protection, we saw a different set of elements involved in price determination. The king's (and/or the marauders') influence on the length of life of the peasant has an effect on the ratio at which the peasant will trade chickens for "insurance" (the chicken-price of insurance). Or, alternatively, the otter hunter who has to pay for hunting rights will sell a more expensive otter skin (higher wampum price of otter skins, or lower otter skin price of wampum), and if people wear otter skins to keep warm they will spend more time chopping firewood and raising sheep when the hunter starts paying hunting rights.

Polanyi's question is not a foolish one. On the contrary, it is an extremely important one and interesting in its own right, but the present thesis is that Polanyi saw different forces, qualitatively different human drives, where the primary differences were in forms.

Polanyi and the Market

Polanyi claims that markets in primitive and archaic societies are (were) qualitatively different from modern markets in societies which are integrated by exchange. He finds two major sources of this difference. Different equivalency (price) formation processes and consequent operation of those processes are posited, which subject has already received treatment here. The second is that supply and demand always represented different sets of individuals in the markets of yore, whereas in modern markets, the same individuals represent both supply and demand.

According to Polanyi's argument, both "supply crowds" and "demand crowds" need not be present simultaneously in the old markets, but they must both be present in the modern market, since they are represented by the same individuals. The argument contends that the combination of supply crowds and demand crowds within the same individuals in modern markets has fooled recent economic theorists into thinking that supply and demand are "combined elemental forces" when, in fact, "each actually consisted of two very different components, namely an amount of goods, on the one hand, and a number of persons, related as buyers and sellers to those goods, on the other" (p. 268).

To persuade the reader that there were, indeed, these two elemental components of supply and demand and that either a supply or a demand crowd could convene a market in the absence of the other, Polanyi gives the example of representatives of a victorious general auctioning off his booty while the general is elsewhere (presumably entertaining himself with his new slaves). In this case, he says, only a demand crowd is present. Nevertheless (so his argument proceeds), the important distinction between goods and people, which this case highlights, is obliterated in ". . . the modern market . . . [in which] there is a price level at which bears turn bulls, and another price level at which the miracle is reversed" (p. 266). In the case of the modern market transaction, one cannot see the single-bladed scissors in action.

This muddle appears to derive from confusion of what demand and supply are (they are quantities "wanted" or "offered" *at particular prices*) with *who* is demanding and *who* is supplying. Polanyi says that the identity of the buyers and sellers in the modern market ". . . has induced many to overlook the fact that buyers and sellers are separate in any other than the modern market . . ." (p. 268). Contrary to the implication here, a person in a modern market is not buying from and selling to himself. That is nonsense. If Polanyi is thinking of a broker of some sort who acts as agent for different individuals who possess different goods, the same broker may buy a good *from* one of his patrons and sell it *to* the other, but this is a different case. The individuals actually concerned in the transaction—that is, the one who is demanding and the one who is supplying—are different, even though their agents are one and the same; and it is Polanyi who has been deluded. Let us try to make sense of this idea of market, and see if what happens in a modern market might happen in a nonmodern market, too.

In current economic terminology, Polanyi is toying with the concept of excess demand or excess supply: at a certain price (say, p_o) for a particular item, some individuals may possess more of that item than they wish to hold; (excess supply = negative excess demand); others may find that they have less than they want, (excess demand = negative excess supply) if p_o is the price. At some price, the sum of excess demands (excess supplies) will be zero, implying that, given the new price, everyone has just the amount he wants. (This is not to be confused with the possibility that some person still wants some more, but cannot afford it. He has the choice of spending more of his income to obtain more of the item (say, good Y), but at the current price—measured in terms of say, good X—he chooses to not give up any more of good X to get good Y. Polanyi has first erred by saying that some price *level* will change buyers into sellers, and vice versa. If the price level moves up or down, there is no implication for *relative* prices of individual goods—for example, the price of sugar in terms of milk. Only relative price changes will perform this miracle. Now, perusing the *relative* price changes and granting that this "bull-into-bear" phenomenon occurs in the modern market, is it a useful concept for nonmodern societies?

Again, let us pose a situation. Imagine a society which eats two foods, fish and corn. One group of these people devotes their time to growing corn, and the other group fishes, and they trade their respective harvests. If the corn growers are particularly interested in eating corn, the exchange ratio of fish for corn will rise from, say, p_o to p_1. At the relative price p_o, there was excess demand for corn; people who formerly sold corn now sell less, and some fishermen will now decide to get rid of some of the corn they had bought earlier, since they can now get more fish for it. Or consider the same society, after a storm wipes out the corn crop. Corn farmers start fishing, since corn will not grow till next year and they have nothing else to offer for fish. The desire to obtain fish (the demand for it) increases (because there is nothing else to eat), and people who formerly bought it now go fishing and supply it themselves. That they supply for themselves makes no difference; the new value of fish sends them into that endeavor called "supplying fish."

Inasmuch as Polanyi's analysis of prices and markets has relied on myopic observations, suppositions which fly in the face of common sense, and definitional non sequiturs, any resulting outlook would view the world through an upside-down, distorting mirror. The distinctions made between types of equivalency ratios have depended upon other unstated things remaining constant. His insistence upon the separation of demand crowds and supply crowds in the concept of market has confused the people transacting and the items transacted, among other egregious features. And he has defined a modern market so that when the price is a certain amount, a man buys from himself, and when it is higher, he sells to himself. The entire apparatus is so confusing that it is amazing that it has been accepted nearly as gospel by a large number of scholars.

II. ECONOMICS OF PRIMITIVE EXCHANGE

Adherence to Polanyi's classification of exchange types, in order to capture some theoretical explanation (or interpretation) for the interpenetration of economic activity with social custom in primitive societies, has restricted the economic geography of those societies largely to description. Nevertheless, the economic anthropology which has its roots in Polanyi's scheme, and upon which this geographic literature rests, need not remain on such a level.

Though conceived within the substantivist tradition, Sahlins's work on exchange in primitive societies contains the seeds of much good economic reasoning (Sahlins, 1965a;

1965b; 1972, chaps. 5–6). His inherited tendency to shy away from economic theory, however, combined with the obvious institutional differences between the societies under study and the one of which Sahlins is a member himself, halts him short of a powerful *and* simple theory. Sahlins applies the exchange-mode framework (reciprocity, redistribution, and exchange) to the observation that different exchange ratios (prices) are in effect between different subsets of individuals. Sahlins's main points will be presented and a more generalized explanation developed, which, it is contended, will generate more information with the same amount of information input.

Variations in Exchange Ratios and Reciprocity

Sahlins presents the observation that exchange ratios of particular goods vary systematically with kinship distance, relative rank between parties, relative wealth between parties, and physical distance. He suggests the existence of different kinship groups, each such group being contained within an inclusive, larger group. Thus the individual would be at the center; the immediate group containing himself and similar groups (that is, other individuals) may be his own (or patrilineal, matrilineal, and so forth) family; containing his own family and several related families may be a larger kinship, or tribal, group; and so on, forming concentric circles in the extent of kinship enclosure. Exchanges occur between persons, both of whom are within the groups closer to the center, in a very casual manner, often (casually) without consideration of "balance,"—that is, whether the transfer of an item from one person to another will be reciprocated with some item (items) of approximate value, or within any particular time period. But if one of these individuals engages in a transaction with a person who gains inclusion into the larger group of which the first person is a member only in one of the farther "concentric circles" (that is, he is more distantly related), the exchange is likely to be more closely accounted. Some reciprocation will be expected, and a more definite (or shorter) time period is implied (or explicit). In exchanges with persons of still more distant kinship, accounting will become more rigid, and, implicitly or explicitly, exchange ratios will increase (that is, more of the other individual's item will be required to adequately compensate the first individual for the item he offers). The same phenomenon is found with physical distance, also, though the relation between physical and kinship distance may be close.

Holding kinship and physical distance constant, the greater the disparity in rank between two transacting parties, the more likely is one to find exchange ratios affected. Whether the ratio moves in favor of the higher or lower ranked individual depends, as Sahlins says, "on whether he [the big man] is developing rank or capitalizing on it" (1965, p. 98). Holding all three of these factors constant, differences in wealth are likely to shift exchange ratios in favor of the poorer individual.

Sahlins explains the phenomenon of the kinship distance-exchange ratio covariation within Polanyi's exchange-mode framework by developing a three-pronged subclassification of reciprocity. Closest to the center of the concentric circle of kinsmen, "generalized reciprocity" occurs; little attention is given by the giver (or by the receiver) to considerations of what is to be returned and when. "Balanced reciprocity" characterizes exchanges between more distant kinsmen. The spirit is such that close accounting is kept, but transacting parties are concerned for the resultant satisfaction of each other. At greatest distances (possibly outside the extent of any "circle"), "negative reciprocity" occurs, with each party trying to get something for nothing. Characteristic of societies in which generalized reciprocity accounts for a large proportion of transactions is particular closeness (solidarity) of social units; for example, of extended families, tribes, villages. Contrasting this more unified appearance is the atomistic character of social units when balanced reciprocity dominates. Sahlins limits negative reciprocity to transactions between very widely separated (measured in kinship or physical distance) individuals.

The example that Sahlins gives of balanced reciprocity involving trade in staple products between primitive, Southeast Asian societies and the more advanced urban center illustrates and, to some extent, rationalizes the association of social atomicity with balanced reciprocity. According to Sahlins's explanation, the availability of the urban market for rice requires that individual households be permitted to accumulate rice. If social rules required that more fortunate (more attentive?) households freely distributed rice to less fortunate households, the trade relations could wither, as less and less rice was left over for trade. Consequently, to sustain the trade, the largest group with an identified self-interest is the household, and sharing is not a widely practiced social virtue.

Sahlins's reaction to the existence of this situation is:

> In three southeast Asian communities the prevalence of balanced reciprocity does seem connected with special circumstances. But then the circumstances suggest that it is not legitimate to involve these people in the present context of tribal economics. . . . Perhaps they are but classed with peasants . . . (1965b, pp. 181–82)

It appears that Sahlins's retread of Polanyi's exchange-type framework has run up against a brick wall. Unfortunately, Sahlins seems to abdicate the effort to relate the occurrence of sharing to its nonoccurrence, although his scheme of subdividing reciprocity into a continuum implies the existence (or at least Sahlins's supposition of the existence) of a common set of forces. Let us now turn to a simplified unification of Sahlins's observations, explanations, and abdications.

Interdependent Utilities

Sahlins acknowledges the existence of some set of satisfactions, which we will call utility, and which is derived from and is the objective of transactions. The ratio of exchange between units of two different items (which we will now call "price") is determined by interactions between the preference systems (utility functions) of the individuals involved and the amount of the commodities in existence. Ordinarily, the utility function of, say, individual i is formulated with its only arguments being the consumption of particular commodities by individual i. But most individuals are concerned with the well-being of persons other than themselves, either for the better or for the worse. Consequently, one will be happy (that is, be in equilibrium in his consumption habits) when the ratio of the increments in his satisfaction, contributed by (1) his own consumption and (2) his neighbor's well-being, is equal to the relative cost to him, in terms of his own contribution to his neighbor's consumption. The usual economic formulation of this is $MU_{wj}/MU_c = P_w/P_c$, where MU_{wj} is the increment in i's satisfaction coming from an increase in his neighbor's well-being (i's marginal utility of j's well-being); MU_c is the increment in i's satisfaction, contributed by an increase in his own consumption; and using p_c, the price of his own consumption, as a numeraire, intragroup transfers of resources to produce well-being in others will continue as long as individual i's subjective value of seeing his kinsman better off exceeds the objective cost to him of doing so. In such manner one can explain economically (and predict) the existence of such intragroup transfers at expected returns to the giver of less than the resource cost.

This scheme can also predict such "atomization" of communities, such as occurred in the Southeast Asian case, cited above, in which the opportunities for trade in a food crop (rice) eliminated most (if not all) forms of extra-household sharing. If the price of rice increases as the opening of trade with the advanced urban centers puts the tribal society in contact with a greater demand for rice, some rearrangement of rice allocation among its several alternative uses will increase the satisfaction of the individual tribesmen. Unless the preference for his neighbor's well-being (as reflected in his rice consumption) increases *pari passu* with the increase in the price of rice, sharing will decline. If the price of rice goes high enough, it will cease altogether between households, and, as Sahlins observes, will induce individuals within a single household to ration it among themselves more judiciously. In this case, the number of exchange ratios between rice and other items will decline, and one may, indeed, find the only nonmarket price exchange ratio (transfer price) for rice between members of the same household.

Accounting Units

This brings us to the issue of the accounting unit. In ordinary consumer theory, the individual is the most frequent (and always the ultimate) unit of account. It is the individual who has preferences, faces prices, experiences a budget constraint in the form of his income plus whatever he can borrow, and must make his allocation decisions with reference to his own resource limitations.

In a firm, when an item is transferred from one branch of the firm to another, no sale occurs, but since the firm is shifting the allocation of real resources, there is a cost to the sending branch and a benefit to the receiving branch. The question arises as to what value should be assigned to the item transferred, since the firm headquarters would be foolish to transfer an item from one office to another if that item was more productive in its first use than in the second. The theory of the firm indicates that the headquarters should value the transferred item at its market price to maximize its return from that resource only under conditions of perfect competition. Given that market price, the firm has three options for the item's use: use it in branch 1; transfer it to branch 2 and use it there; or sell it outside the firm. The cost to branch 1 must be less than the return to branch 2, which must, in turn, be greater than or equal to the market value of the item. There are certain similarities between the hierarchical structure of a firm and the organization of the sorts of social groupings that comprise Sahlins's concentric circles of kinship. There is a head in the social grouping, comparable to the firm headquarters. Within the firm are branches or offices, while individuals, or households, are the members of the social group. For both structures, there is a boundary separating the inside from the outside, and a balance of payments between inside and outside.

Let us consider the objectives of the two organizations. The firm maximizes profits (let us assume): total receipts minus total costs. Suppose the social group maximizes its group utility, which is some combination of the utilities of the individual members. If the utilities of the members are interdependent—that is, if each individual's utility function contains the utility of his neighbor(s) as an argument—additivity is not possible. There are several alternative formulations. The group leader can establish the group's preference function and impose it on everyone else (the dictatorial solution). As an explanation of reality, this does not seem particularly plausible. Suppose, instead, that each individual maximizes his own utility, which contains others' utilities as arguments, and that this maximization is subject to some modified budget constraint *and* to some socially accepted distribution of wealth or income.

The ordinary conception of the individual's budget constraint, without borrowing, implicitly gives him a zero (balanced) balance of payments. If the relevant accounting unit is the group rather than the individual, the individual will maximize his satisfactions subject to a budget constraint which may be more or less than his own produced income. Transfers between individuals within the group will not be considered as surpluses or deficits on their personal balances of payments, since only the group keeps such accounts. Thus a series of intragroup transfers can increase the utility of all involved without the appearance of accounted debt. In this scheme, however, it should be noted that the budget modification is not an ad hoc rearrangement to fit the circumstances; instead, it follows from the interdependence of utilities. We will now present this theory more formally.

Formal Presentation of the Theory

Consider an individual, i, who has a utility function, U_{ij} which depends on his consumption of a composite bundle of commodities, X_j, and the utility of his neighbor, U_j:

$$U_i = f_i(x_i, U_j), \frac{\partial f_i}{\partial x_i} > 0, \frac{\partial f_i}{\partial U_j} > 0, \frac{\partial^2 f_i}{\partial x_i \partial U_j} > 0, \frac{\partial^2 f_i}{\partial x_i^2} < 0, \frac{\partial^2 f_i}{\partial U_j^2} < 0. \quad (1)$$

To maximize his utility, he maximizes his consumption while keeping an eye on his neighbor's well-being, all subject to a budget constraint. His object is thus to

$$\text{Maximize } U_i = f(x_{ij}, U_j) \quad (2)$$

$$\text{subject to } p_x x_i + p_t t_{ij} = I_i \ .$$

p_x is the price of the composite commodity x. t_{ij} is a transfer of some subset of commodities from individual i to individual j, chosen so that p_x does not change (one might suppose that t_{ij} is some exchange commodity which may be somehow converted back into commodity x), and p_t is the price of this group of commodities. I_j is i's budget constraint.

Individual j performs the same sort of operation:

$$\text{Max } U_j = f_j(x_j, U_i) \quad (3)$$

$$\text{s.t. } p_x x_j + p_t t_{ji} = I_j$$

The transferred commodities term in i's budget constraint, t_{ij}, is measured positively, indicating that the value of his consumption of x is less than his total budget. In j's budget constraint, the transfer term is written t_{ji} and is measured negatively, indicating that the transfer permits him to consume a value greater than his income (that is, his budget). Without working through the usual budget-constrained utility maximization exercise, several points made previously regarding the economic explanation of transfers and variable exchange ratios can be demonstrated with some simple, algebraic manipulations.

The behavioral relation between the transfer in i's budget constraint and j's utility is $dt_{ij}/dU_j > 0$: when j's utility (consumption of x_j) is lower, the transfer from i to j is larger positive.

It is obvious from equation (2) that i's consumption of x_i falls as he transfers some of his income to j. Differentiating the budget constraint totally, and keeping prices and income constant:

$$p_x dx_i + p_t dt_{ij} = dI_i = 0 \quad (4)$$

$$\frac{dx_i}{dt_{ij}} = -\frac{p_t}{p_x} < 0 \ .$$

A decrease in income to j reduces utility of both individuals, but the fall in the utilities need not be proportional. Differentiate j's budget constraint totally, keeping prices constant:

$$p_x dx_j + p_t dt_{ij} = dI_j \tag{5}$$

$$p_x \frac{dx_j}{dI_j} = 1 - p_t \frac{dt_{ji}}{dI_j}$$

and $dx_j/dI_j > 0$ in the absence of a transfer. So, for $dI_j < 0$ representing a decrease in income for individual j,

$$\frac{\partial f_i}{\partial x_j} \frac{dx_j}{dI_j} < 0 \ .$$

Putting this into i's utility function,

$$\frac{\partial f_i}{\partial U_j} dU_j < 0$$

for $dI_j < 0$. Some transfer $t_{ij} > 0$ ($t_{ji} < 0$) will occur to maximize U_i and U_j subject to their new interrelated constraint, j's lower income. If j's consumption of x is to remain unchanged, an incremental transfer from i to j of

$$\frac{dt_{ij}}{dI_j} = -\frac{1}{p_t}$$

is required. But this transfer will reduce i's consumption of x, and hence i's utility, and this will prevent the restoration of j's consumption of x from fully restoring j's utility.

To see this, differentiate i's budget constraint totally, substitute the above value of the transfer required to keep j's consumption constant, and put the resulting decrease in i's consumption into i's totally differentiated utility function.

$$p_x dx_i + p_t dt_{ij} = dI_i = 0 \tag{6}$$

for a change in j's income of dI_j,

$$p_x \frac{dx_i}{dI_j} + p_t \frac{dt_{ij}}{dI_j} = 0 \ . \tag{7}$$

Using $dt_{ij} = -dt_{ji}$ and $\dfrac{dt_{ji}}{dI_j} = \dfrac{1}{p_t}$,

$$p_x \frac{dx_i}{dI_j} = 1 \ , \text{ and} \tag{8}$$

the decrease in i's consumption consequent upon the effort to compensate j is

$$dx_i = \frac{dI_j}{p_x} < 0 \ (\text{for } dI_j < 0) \ . \tag{9}$$

Substituting this value of dx_j into i's utility function,

$$dU_i = \frac{\partial f_i}{\partial x_i}\left(\frac{dI_j}{p_x}\right) + \frac{\partial f_i}{\partial U_j} dU_j \ . \tag{10}$$

The first term on the right hand side of the last equation is negative, but dU_i cannot be determined without knowing dU_j. But we have kept j's consumption of x constant, so in j's utility function, the first term on the right side is zero:

$$dU_j = \frac{\partial f_j}{\partial x_j} dx_j + \frac{\partial f_j}{\partial U_i} dU_i \tag{11}$$

But this gives us two equations in two unknowns (dU_i and Du_j), and we can solve for the final state of satisfactions of two individuals. In matrix notation, the equations are

$$\begin{pmatrix} 1 & -\frac{\partial f_i}{\partial U_j} \\ -\frac{\partial f_j}{\partial U_i} & 1 \end{pmatrix} \begin{pmatrix} dU_i \\ dU_j \end{pmatrix} = \begin{pmatrix} \frac{\partial f_i}{\partial x_i}\left(\frac{dI_j}{p_x}\right) \\ 0 \end{pmatrix} \qquad (12)$$

The resulting changes in utility for the two individuals are

$$dU_i \Big|_{dx_j=0} = \frac{\partial f_i}{\partial x_i}\left(\frac{dI_j}{p_x}\right) \cdot \frac{1}{\Delta} < 0$$

and

$$dU_j \Big|_{dx_j=0} = \frac{\partial f_j}{\partial U_i} \cdot \frac{\partial f_i}{\partial x_i}\left(\frac{dI_i}{p_x}\right) \cdot \frac{1}{\Delta} < 0 ,$$

where

$$\Delta = 1 - \frac{\partial f_j}{\partial U_i} \cdot \frac{\partial f_i}{\partial U_j} > 0 \left(\text{assuming} \frac{\partial f_j}{\partial U_i} \text{ and } \frac{\partial f_i}{\partial U_j} < 1\right) .$$

At this point let us digress a moment on the ratio of changes in utility for i and for j. Although the partial derivative of each utility function with respect to the neighbor's well-being

$$\left(\frac{\partial f_i}{\partial U_j}, \frac{\partial f_j}{\partial U_i}\right)$$

is positive, the total derivative of the change in one individual's utility with respect to a change in the other's utility may be negative. A negative rate of transformation of one individual's utility into the other's would imply that adversary relations will still exist in such a society. In the present formulation, however, this point is especially emphasized by the construction of the problem: there are only two parties, and exclusivity of resources implies that any material gain for one is necessarily a material loss for the other. This presents an interesting contrast between the nonmaterial rate of transformation of utilities, which may be positive or negative, and the trade-off between one person's utility and the other person's consumption (remembering that $dx_j = -dx_i$), which is strictly negative. If additional parties were introduced (foreigners, for instance), it is possible that strict negativity would obtain for the utility transformation rate between some parties.

The derivation of this rate of transformation of satisfaction for individual i into satisfaction for individual j is as follows. Differentiating equations (2) and (3) totally, we have the relationships contained in equations (10) and (11).

$$dU_i = \frac{\partial f_i}{\partial x_i} dx_i + \frac{\partial f_i}{\partial U_j} dU_j \qquad (10.A)$$

$$dU_j = \frac{\partial f_j}{\partial x_j} dx_j + \frac{\partial f_j}{\partial U_i} dU_i \qquad (11.A)$$

Setting $dx_j = dx_i$, since we have postulated a two-individual world, dividing (10.A) by dU_j and (11.A) by dU_i to obtain rate-of-transformation forms, and rearranging, we obtain

$$\frac{dU_i}{dU_j} = \frac{\partial f_i/\partial x_i}{\partial U_j/dx_i} + \frac{\partial f_i}{\partial U_j} \qquad (10.B)$$

$$\frac{-dU_j}{dx_i} = \frac{\partial f_j}{\partial x_j} + \frac{\partial f_j}{\partial U_i}\ . \qquad (11.B)$$

Forming the system

$$\begin{pmatrix} 1 & -\dfrac{\partial f_i}{\partial x_i} \\ & \\ 0 & -1 \end{pmatrix} \begin{pmatrix} \dfrac{dU_i}{dU_j} \\ \\ \left(\dfrac{dU_j}{dx_i}\right)^{-1} \end{pmatrix} = \begin{pmatrix} \dfrac{\partial f_i}{\partial U_j} \\ \\ \left(\dfrac{\partial f_j}{\partial x_j} + \dfrac{\partial f_j}{\partial U_i}\right)^{-1} \end{pmatrix} \qquad (13)$$

and solving for the rate of transformation, dU_i/dU_j, yields

$$\frac{dU_i}{dU_j} = \frac{\partial f_i}{\partial U_j} - \frac{\partial f_i/\partial x_i}{\left(\dfrac{\partial f_j}{\partial x_j} + \dfrac{\partial f_j}{\partial U_i}\right)} \gtrless 0\ . \qquad (14)$$

Does this represent a concave or convex transformation frontier? The sign of the second derivative of (14) is ambiguous, so the shape of the transformation curve is not clear:

$$\frac{d^2U_i}{dU^2_j} = \frac{\partial^2 f_i}{\partial U_j^2} - \frac{\dfrac{\partial f_i}{\partial x_i}\left(\dfrac{\partial f_i}{\partial U_j} - 1\right)}{\dfrac{\partial f_j}{\partial x_j} + \dfrac{\partial f_j}{\partial U_i}} \gtrless 0\ . \qquad (15)$$

The first term in (15) is negative (see equation 1), but the second term contains the difference between two positive terms, making the total equation indeterminate in sign. (If we continue the assumption used in the determinant of [12], that the partial derivatives of the utility functions are less than unity, the second term is unambiguously positive, but the sign of

$$\frac{d^2U_i}{dU_j^2}$$

still rests on the relative magnitudes of the negative first term and the positive second term.) If we suppose that (15) is negative, yielding the ordinary convex transformation curve, as in figure 5, the frontier will likely have less curvature than a utility transformation curve which plotted independent utilities. This would imply greater substitutability between the satisfactions of individuals in such a society, a characteristic which Sahlins (as well as others writing on gift-giving and wealth redistributing) finds of particular significance in these societies.

Contrasting this complementary effect in utility is the negative relationship between i's consumption of x and j's utility.

$$\frac{dU_j}{dx_i} = -\frac{\partial f_j}{\partial x_j} - \frac{\partial f_j}{\partial U_i} < 0\ . \qquad (16)$$

This gives us the interesting result that the individual may

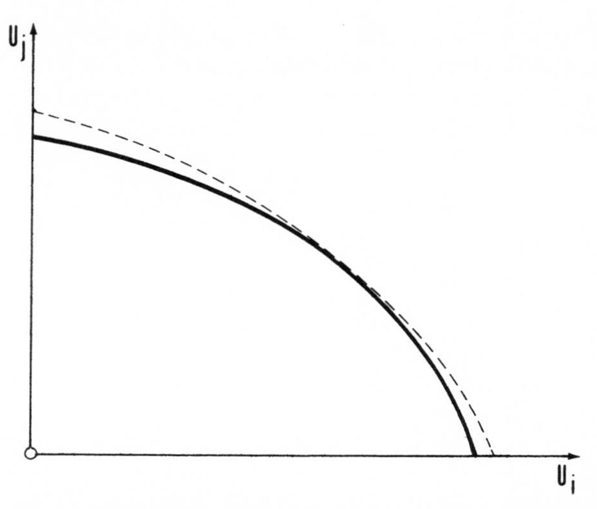

Figure 5. Utility transformation curves with and without interdependent utilities.

actually reduce his utility by increasing his consumption (at his neighbor's expense: because we have specified $dx_j = -dx_i$). If

$$\frac{dU_i}{dU_j} > 0$$

in any region,

$$\frac{dU_i}{dU_j} \cdot \frac{dU_j}{dx_i} = \frac{dU_i}{dx_i} < 0 \ .$$

In such a case, redistribution of consumption will be in the interest of all parties.

Let us now continue with our exercises in redistribution and compensation.

If, rather than fully compensating individual j's consumption at the expense of i's consumption, we wished to divide evenly the change in the utility, we can rearrange equations (10) and (11), totally differentiate the budget constraints, and require that $dU_i = dU_j$. The resulting system of equations is presented below.

$$dU_i = \frac{\partial f_i}{\partial x_i} dx_i + \frac{\partial f_i}{\partial U_j} dU_j \qquad (17)$$

$$dU_j = \frac{\partial f_j}{\partial x_j} dx_j + \frac{\partial f_j}{\partial U_i} dU_i \qquad (18)$$

$$p_x dx_i + p_t dt_{ij} = 0 \qquad (19)$$

$$p_x dx_j - p_t dt_{ij} = dI_j \qquad (20)$$

$$dU_i = dU_j \qquad (21)$$

Solving for the change in transfer required to even the losses in utility, we get

$$dt_{ij} = -\frac{\partial f_j}{\partial x_j}\left(\frac{\partial f_i}{\partial U_j} - 1\right)\frac{dI_j}{\Delta} > 0 \text{ (for } dI_j < 0) \ ,$$

where

$$\Delta = p_t \left\{\frac{\partial f_i}{\partial x_i}\left(\frac{\partial f_j}{\partial U_i} - 1\right) + \frac{\partial f_j}{\partial x_j}\left(\frac{\partial f_i}{\partial U_j} - 1\right)\right\} < 0 \ .$$

This yields the information that: (1) the transfer required will be smaller, the greater is individual i's desire for x consumption relative to individual j's; (2) the transfer is larger, the more important is i's utility in j's utility function

$$\left(\frac{\partial f_j}{\partial U_i} \text{ larger}\right) ;$$

and (3) the transfer is smaller, the more important is j's utility in i's utility function

$$\left(\frac{\partial f_i}{\partial U_j} \text{ larger}\right) .$$

Conclusions (2) and (3) are counter-intuitive, but become understandable when it is considered (for example, in conclusion [3]) that the more sensitive i is to variations in j's welfare, a smaller change in j's welfare will satisfy individual i. Nevertheless, this consequence of the postulate that the well-being of one person enters into the well-being of another person suggests that this postulate could bear some rethinking. According to the formulation, these two individuals are quite sensitive to each other's welfare, but by this definition, though a small decrease in j's welfare will reduce i's welfare considerably, the reverse is symmetrically true: an equally small increase in j's welfare will satisfy i. And, by the specification of concavity for this attribute of neighborly concern (the negative signs of the partial derivatives attributed to the arguments utility in equation [1]), the more j's welfare increases, the less important to i is any further increase. What are the processes of interaction between i and j in jockeying for some optimal income (re-) distribution? As suggested, this issue deserves further consideration, but it will not be presented here.

It can further be demonstrated with this apparatus that variations in prices (p_x and p_t) can also be used to assist the transfer of consumption between individuals. What the model, in this simple version, will not do is yield the result that transfers (or price variations) are more desirable as disparity in wealth increases. A quick look at the utility function should suffice to show that only in the case of fixed total income ($I_i + I_j = \bar{I}$) will changes in the income ratio affect behavior, but this reduces to the case just examined, of an absolute reduction in income to one party. If an argument such as W_i - W_j, for disparity in wealth, were inserted into the utility function, wealth effects on economic behavior would be forthcoming.

Would the geographical study of the economics of nonindustrial societies benefit from a move to a framework such as this? "The day in the life of the predator" might gradually be replaced with "How does the predator decide to spend so much time preying and so much making pots?" Greater emphasis on the reasoning ability of people and fewer appeals to the mental encrustation of habit should permit better insight into the way nonindustrial men organize their territories and translate their environments into their livelihoods.

REFERENCES CITED

Abrams, Ray H. 1943. "Residential Propinquity as a Factor in Marriage Selection, Fifty Year Trends in Philadelphia." *American Sociological Review* 8, no. 3: 288–94.

Ackerman, Edward. 1941. "Sequent Occupance of a Boston Suburban Community." *Economic Geography* 17:51–74.

Allen, W. Bruce. 1972. " 'Gravity Law' of Spatial Interaction: A Comment on the Reply." *Journal of Regional Science* 12, no. 1: 119–26.

Ambrose, P. J. 1968. "An Analysis of Intra-Urban Shopping Patterns." *Town Planning Review* 38, no. 4: 327–34.

Anderson, Theodore R. 1955. "Intermetropolitan Migration: A Comparison of the Hypotheses of Zipf and Stouffer." *American Sociological Review* 20, no. 3: 287–91.

———. 1956. "Potential Models and the Spatial Distribution of Population." *Papers and Proceedings, Regional Science Association* 2:175–82.

Atwood, Wallace A.; Brown, R.; Kniffen, F. B; and Hall, R. B. 1937. "Round Table on Problems in Cultural Geography." *Annals of the Association of American Geographers* 27:155–75.

Barrows, Harlan H. 1962. "Lectures on the Historical Geography of the United States, 1933." Edited by William A. Koelsch. University of Chicago Department of Geography Research Paper, no. 77.

Batty, M. 1970. "Some Problems of Calibrating the Lowry Model," *Environment and Planning* 2:95–114.

——— and Mackie, S. 1972. "The Calibration of Gravity, Entropy, and Related Models of Spatial Interactions," *Environment and Planning* 4:205–33.

Berry, Brian J. L. 1967. *Geography of Market Centers and Retail Distribution.* Englewood Cliffs, N.J.: Prentice-Hall.

———. 1969. *Growth Centers and Their Potentials in the Upper Great Lakes Region.* Washington, D.C.: Upper Great Lakes Regional Commission.

———. 1970a. "Latent Structure of the American Urban System." In B. J. L. Berry, ed. *Classification of Cities: New Methods and Evolving Uses.* San Francisco: Chandler Publishing Co.

———. 1970b. "Hierarchical Diffusion: The Basis of Developmental Filtering and Spread in a System of Growth Centers." In Niles Hansen, ed. *Growth Centers in Regional Economic Development 2* New York: Free Press, 1972.

———. 1971. "City Size and Economic Development: Conceptual Synthesis and Policy Problems, with special reference to South and Southeast Asia." In L. Jacobson and V. Prakash, eds. *Urbanization and National Development* (South and Southeast Asia Urban Studies Annual 1) (Beverly Hills, Calif.: Sage Publications), pp. 111–55.

———; Barnum, Gardiner; and Tennant, Robert J. 1962. "Retail Location and Consumer Behavior." *Papers and Proceedings, Regional Science Association* 9:65–106.

———, and Garrison, William. 1958. "Note on the Central Place Theory and Range of a Good." *Economic Geography* 34:304–11.

———, and Horton, Frank. 1970. *Geographic Perspectives on Urban Systems.* Englewood Cliffs, N.J.: Prentice-Hall.

———, and Wrobel, Andrzej. 1968. *Economic Regionalization and Numerical Methods.* Warsaw: Polish Scientific Publishers.

Bevis, Howard. 1956. "Forecasting Zonal Traffic Volumes." *Traffic Quarterly* 10:207–22.

Bishop, Willard R., Jr., and Brown, Earl H. 1969. "An Analysis of Spatial Shopping Behavior." *Journal of Retailing* 45, no. 2: 23–31.

Black, William R. 1971a. "Substitution and Concentration: An Examination of the Distance Exponent in Gravity Model Community Flow Studies." Discussion Paper Series No. 1. Department of Geography, Indiana University.

———. 1971b. "Utility of the Gravity Model and Estimates of Its Parameter in Commodity Flow Studies." *Papers and Proceedings, Association of American Geographers* 3:28–32.

———. 1972a. "Interregional Commodity Flows: Some Experiments with the Gravity Model." *Journal of Regional Science* 12, no. 1: 107–18.

———. 1972b. "A Comparative Evaluation of Alternative Friction Factors in the Gravity Model." *Professional Geographer* 24, no. 4: 335–37.

Blanchard, R. 1922. "Une Méthode de Géographie Urbaine." *La Vie Urbaine* 4:301–19.

Boas, Franz. 1937. "The Diffusion of Cultural Traits," *Social Research* 4:286–95.

Bobek, H. 1927. "Grundfragen zur Stadtgeographie," *Geographischer Anzeiger* 28:213–24.

———. 1928. "Innsbruck: eine Gebirgsstadt, ihr Lebensraum und ihre Erscheinung." *Forschungen Zur Deutschen Landes—Und Volkskunde* 25:220–372.

———. 1936. "Eine neue Arbeit zur Stadtgeographie." *Zeitschrift Der Gesellschaft Fur Erdkunde Zu Berlin,* pp. 125–30.

Boon, Francoise. 1967. A *Simple Model for the Diffusion of an Innovation in an Urban System.* Chicago: Center for Urban Studies, University of Chicago.

Borts, George H. 1967. *Patterns of Regional Economic Development in the United States.* Report to the National Advisory Commission on Rural Poverty, U.S. Department of Agriculture.

Bossard, James H. 1932. "Residential Propinquity as a Factor in Marriage Selection." *American Journal of Sociology* 38, no. 2: 219–24.

Bowers, Raymond V. 1937. "The Direction of Intra-Societal Diffusion." *American Sociological Review* 2:826–36.

Bridgman, Percy Williams. 1922. *Dimensional Analysis.* New Haven: Yale University Press.

Bright, Margaret L., and Thomas, Dorothy S. 1941. "Interstate Migration and Intervening Opportunities." *American Sociological Review* 6, no. 6: 773–83.

Britton, John N. H. 1971. "The Interaction Model and Relative Location Variables." *Geografiska Annaler* 53B, no. 2: 129–37.

Broeck, Jan O. M. 1932. *The Santa Clara Valley, California: A Study in Landscape Changes.* Utrecht: Oosthoek.

Brown, Lawrence A. 1968a. *Diffusion Dynamics.* Lund Studies in Geography, Series B, Human Geography, no. 29.

———. 1968b. *Diffusion Processes and Location.* Philadelphia: Regional Science Research Institute, Bibliography Series, no. 4.

———, and Horton, Frank E. 1970. "Functional Distance: An Operational Approach." *Geographical Analysis* 2, no. 1: 76–82.

Brown, Ralph H. 1943. *Mirror for Americans: Likeness of the Eastern Seaboard, 1810.* New York: American Geographical Society.

———. 1948. *Historical Geography of the United States.* New York: Harcourt, Brace.

Bucklin, Louis P. 1971a. "Trade Area Boundaries: Some Issues in Theory and Methodology." *Journal of Marketing Research* 8:30–37.

———. 1971b. "Retail Gravity Models and Consumer Choice: Theoretical and Empirical Critique." *Economic Geography* 47, no. 4: 489–97.

Bunge, William. 1966. *Theoretical Geography.* Lund Studies in Geography Series C, General and Mathematical Geography no. 1. Lund, Sweden: C. W. K. Gleerup.

Cancian, Frank. 1965. *Economics and Prestige in a Maya Community: The Religious Cargo System in Zinacantan.* Stanford: Stanford University Press.

Carey, H. C. 1858. *Principles of Social Science.* Philadelphia: J. B. Lippincott and Co.

Carlberg, B. 1926. "Stadtgeographie." *Geographischer Anzeiger* 27:148–53.

Carlson, Carl F. 1940. "Aurora, Illinois: A Study in Sequent Land Use." Ph.D. diss., University of Chicago.

Carroll, J. Douglas. 1955a. "Defining Urban Trade Areas." *Traffic Quarterly* 9, no. 2: 149–61.

———. 1955b. "Spatial Interaction and the Urban-Metropolitan Regional Description." *Papers and Proceedings, Regional Science Association* 1:D1-D14.

———, and Bevis, Howard W. 1957. "Predicting Local Travel in Urban

Regions." *Papers and Proceedings, Regional Science Association* 3:183–97.

Carrothers, Gerald A. P. 1956. "Historical Review of the Gravity and Potential Concepts of Human Interaction." *Journal of The American Institute of Planners* 22:94–102.

———. 1958. "Population Projection by Means of Income Potential Models." *Papers and Proceedings, Regional Science Association* 4:121–52.

Casetti, E., and Semple, R. K. 1969. "Concerning the Testing of Spatial Diffusion Hypotheses." *Geographical Analysis* 1:254–59.

Cavanaugh, Joseph A. 1950. "Formulation, Analysis and Testing of the Interactance Hypothesis." *American Sociological Review* 15, no. 5: 763–66.

Cesario, F. J. 1971. "Parameter Estimation in Trip Distribution Modeling," Ph.D. diss., Ohio State University.

Chandler, William R. 1948. "The Relationship of Distance to the Occurence of Pedestrian Accidents." *Sociometry* 11, no. 1: 108–10.

Chicago Area Transportation Study. 1959. *Survey Findings.* Chicago: Chicago Area Transportation Study.

Chinitz, Benjamin. 1956. "Discussion: Population Project by Means of Income Potential Models." *Papers and Proceedings, Regional Science Association* 4:159–60.

Chorley, Richard J., and Haggett, P. 1967. *Models in Geography.* London: Methuen and Co.

Christaller, W. 1933. *Die Zentralen Orte in Süddeutschland.* Jena: Fischer. Translated into English by C. W. Baskin. 1966. *Central Places in Southern Germany.* Englewood Cliffs, N.J.: Prentice-Hall.

———. 1938. "Rapports Fonctionels entre les Agglomerations Urbaines et les Campagnes." *Congrés International de Géographie, Comptes Rendus, Section 3a* (2): 123–38.

Christensen, David E. 1956. "Rural Occupance in Transition: Lee and Sumter Counties, Georgia." University of Chicago, Department of Geography Research Paper No. 43.

Claeson, Frederik. 1968. "Distance and Human Interaction." *Geografiska Annaler,* Series B—Human Geography 50B (2):142–61.

Clark, Andrew H. 1954. "Historical Geography." In P. E. James and C. F. Jones, eds. *American Geography: Inventory and Prospect.* Syracuse, N.Y.: Syracuse University Press, pp. 71–105.

———. 1960. "Geographical Change: A Theme for Economic History." *Journal of Economic History* 20:607–13.

Clark, Colin, and Peters, G. H. 1965. "The Intervening Opportunities Method of Traffic Analysis." *Traffic Quarterly* 19:104–15.

Clark, W. A. 1968. "Consumer Travel Patterns and the Concept of Ranges." *Annals of the Association of American Geographers* 58, no. 2: 386–96.

Clausius, R. J. 1885. *Die Potenlealfunction und das Potential.* Leipsig: J. A. Barth.

———. 1899. *The Second Law of Thermodynamics: Memoirs of Cornot, Clausius and Thompson.* New York: Hocker and Bros.

Converse, P. D. 1949. "New Laws of Retail Gravitation." *Journal of Marketing* 14, no. 3: 379–90.

Crain, Robert L. 1966. "Fluoridation: The Diffusion of an Innovation Among Cities." *Social Forces* 44:467–76.

Curry, James P. 1970. *Updating the Opportunity Model for Continuing Transportation Planning.* Chicago: Chicago Transportation Study.

———. 1972. "Spatial Analysis of Gravity Flows." *Regional Studies* 6:131–47.

Dalton, George. 1969. "Theoretical Issues in Economic Anthropology." *Current Anthropology* 10:63–80.

Darby, H. C. 1960. "An Historical Geography of England Twenty Years After." *Geographical Journal* 126:147–59.

———, ed. 1973. *A New Historical Geography of England.* Cambridge: Cambridge University Press.

Davie, Maurice R., and Reeves, Ruby Jo. 1939. "Propinquity of Residence Before Marriage." *American Journal of Sociology* 44, no. 4: 510–17.

Davis, Charles M. 1936. "The High Plains of Michigan." *Michigan Papers in Geography* 6:303–41.

Davis, William Morris. 1906. "An Inductive Study of the Content of Geography." *Bulletin of the American Geographical Society* 38:67–84.

De La Blache, Vidal P. 1926. *Principles of Human Geography.* Edited by Emmanuel De Martonne. New York: Henry Holt and Co.

Detroit Metropolitan Area Traffic Study. 1955. *Data Summary and Interpretation.* Detroit: Detroit Metropolitan Area Traffic Study.

———. 1956. *Future Traffic and a Long Range Expressway Plan.* Detroit: Detroit Metropolitan Area Traffic Study.

Dillman, C. Daniel. 1971. "Occupance Phases of the Lower Rio Grande of Texas and Tamaulipas." *California Geographer* 12:3–67.

Dodd, Stuart C. 1936a. "The Standard Error of the Estimate." *Annals of Mathematical Statistics* 7, no. 4: 202–9.

———. 1936b. "A Theory for the Measurement of Some Social Forces." *Scientific Monthly* 63:58–62.

———. 1942. *Dimensions of Society: A Qualitative Systematics for the Social Sciences.* New York: Macmillan.

———. 1943a. "Of What Use Is Dimensional Sociology?" *Social Forces* 22, no. 1: 169–82.

———. 1943b. "Induction, Deduction and Causation." *Sociometry* 5, no. 2: 119–49.

———. 1943c. "Sociometry Delimited: Its Relation to Social Work, Sociology, and the Social Sciences." *Sociometry* 6, no. 3: 200–205.

———. 1943d. "Operational Definitions." *American Journal of Sociology* 48, no. 4: 482–91.

———. 1950. "The Interactance Hypothesis—A Gravity Model Fitting Physical Masses and Human Groups." *American Sociological Review* 15, no. 1: 245–57.

———. 1951a. "Sociomatrices and Levels of Interaction." *Sociometry* 14, no. 2: 237–47.

———. 1951b. "Scientific Methods in Human Relations." *American Journal of Economics and Sociology* 10, no. 3: 221–36.

———. 1955. "Diffusion Is Predictable: Testing Probability Models for Laws of Interaction." *American Sociological Review* 20, no. 4: 392–401.

Dodge, Richard E. 1938. "The Interpretation of Sequent Occupance." *Annals of the Association of American Geographers* 28:233–37.

Dodge, Stanley D. 1931. "Sequent Occupance on an Illinois Prairie." *Bulletin of the Geographical Society of Philadelphia* 29:205–9.

Dörries, H. 1930. "Der gegenwärtige Stand der Stadtegeographie." *Petermanns Geographische Mitteilungen, Ergänzungsband* 209:310–25.

Duncan, William J. 1953. *Physical Similarity and Dimensional Analysis: An Elementary Treatise.* London: E. Arnold.

Dunn, Edgar S. 1956. "The Market Potential Concept and the Analysis of Location." *Papers and Proceedings, Regional Science Association* 2:183–94.

Edens, H. J. 1970. "Analysis of a Modified Gravity Model." *Transportation Research* 4:51–62.

Eilon, S.; Fowkes, T. R.; and Tilley, R. P. R. 1969. "Analysis of a Gravity Demand Model." *Regional Studies* 3, no. 2: 115–22.

Eiselen, Elizabeth. 1943. *A Geographic Traverse Across South Dakota.* Ph.D. diss., University of Chicago.

Ellis, Jack, and Van Doren, Carlton. 1966. "Comparative Evolution of Gravity and System Theory Models for Statewide Recreational Traffic Flows." *Journal of Regional Science* 6, no. 2: 57–70.

Ericksen, Sheldon D. 1953. "Occupance in the Upper Deschutes Basin, Oregon." University of Chicago, Department of Geography Research Paper No. 32.

Espenshade, Edward B., Jr. 1944. "Urban Development at the Upper Rapids of the Mississippi." Ph.D. diss., University of Chicago.

Evans, A. W. 1970. "Some Properties of Trip Distribution Methods." *Transportation Research* 4:19–36.

———. 1971. "The Calibration of Trip Distribution Models with Exponential or Similar Cost Functions." *Transportation Research* 5:15–38.

Febvre, Lucien, with Batillon, Lionel. 1925. *A Geographical Introduction to History.* New York: Alfred A. Knopf.

Focker, Charles M. 1953. *Dimensional Methods and Their Applications.* London: E. Arnold.

Folger, John. 1953. "Some Aspects of Migration in the Tennessee Valley." *American Sociological Review* 18:253–60.
Friedmann, John. 1966. *Regional Development Policy.* Cambridge, Mass.: M.I.T. Press.
———, and Miller, John. 1965. "The Urban Field," *Journal of the American Institute of Planners* 31:312–19.
Garrison, William L. 1955. "The Spatial Impact of Transport Media." *Papers and Proceedings, Regional Science Association* 1:T1–T14.
———. 1956. "Estimates of the Parameters of Spatial Interaction." *Papers and Proceedings, Regional Science Association* 2:280–88.
Geisler, W. 1920. "Beiträge zur Stadtgeographie." *Zeitschrift Der Gesellschaft Für Erdkunde Zu Berlin,* pp. 274–97.
———. 1924. "Die deutsche Stadt: ein Beitrag zur Morphologie der Kulturlandschaft." *Forschungen Zur Deutschen Landes—Und Volkskunde* 22, no. 5: 359–543.
Gilfillan, S. C. 1920. "The Coldward Course of Progress." *Political Science Quarterly* 35:393–99.
Glacken, Clarence J. 1967. *Traces on the Rhodian Shore.* Berkeley: University of California Press.
Glasgow, James. 1939. *Muskegon, Michigan: The Evolution of a Lake Port.* Ph.D. diss., University of Chicago.
Glejser, H., and Dramis, A. 1969. "A Gravity Model of Interdependent Equations to Estimate Flow Creation and Diversion." *Journal of Regional Science* 9, no. 3: 439–50.
Glendinning, Robert M. 1935. "The Lake St. Jean Lowland, Province of Quebec." *Michigan Papers in Geography* 5:313–41.
Golledge, R. G. 1970. "Some Equilibrium Models of Consumer Behavior." *Economic Geography* 42, no. 2, Supplement: 417–24.
Gould, Peter. 1969. *Spatial Diffusion, Commission on College Geography Resource Paper,* no. 4. Washington, D.C.: Association of American Geographers.
———. 1972. "Pedogogic Review." *Annals of the Association of American Geographers* 62, no. 4: 689–700.
Gradmann, R. 1901. "Das mitteleuropäische Landschaftsbild nach seiner landschaftlichen Entwicklung." *Zeitschrift Der Gesellschaft Für Erdkunde Zu Berlin,* pp. 361–77, 435–47.
———. 1913a. "Das ländliche Siedlungswesen des Königreichs Wurttemberg," *Forschungen Zur Deutschen Landesund Volkskunde* 21, no. 1: 1–136.
———. 1913b. "Die städtischen Siedlungen des Königreichs Württemberg." *Forschungen Zur Deutschen Landesund Volkskunde* 21, no. 2: 137–226.
———. 1916. "Schwäbische Städte." *Zeitschrift Der Gesellschaft Für Erdkunde Zu Berlin,* pp. 425–57.
———. 1937. "Zur siedlungsgeographischen Methode." *Geographische Zeitschrift* 43:353–61.
Graebner, Fritz. 1911. *Methode de Ethnologie.* Heidelberg: Carl Winter.
Gras, N. S. B. 1922. "The Development of Metropolitan Economy in Europe and America." *American Historical Review* 27:695–708.
Green, Howard L. 1956. "Discussion: Gravity and Potential Models of Spatial Interaction." *Papers and Proceedings, Regional Science Association* 2:195.
Gregor, Howard F. 1953. "Agricultural Shifts in the Ventura Lowland of California." *Economic Geography* 29:340–61.
Gukhaman, Aleksandr A. 1965. *Introduction to the Theory of Similarity.* New York: Academic Press.
Gulley, J. L. M. 1959. "The Turnerian Frontier: A Study in the Migration of Ideas." *Tijdschrift Voor Economische en Sociale Geografie* 50:65–72, 81–91.
Hägerstrand, Torsten. 1953, 1967. *Innovation Diffusion as a Spatial Process.* Chicago: University of Chicago Press.
Haggett, Peter. 1965. *Locational Analysis in Human Geography.* London: Edward Arnold.
———, and Chorley, Richard J. 1969. *Network Analysis in Geography.* London: Edward Arnold.
Hall, Robert B. 1935. "The Geographical Region: A Resume." *Annals of*

the Association of American Geographers 25, no. 3: 122–36.
———. 1937. "Tokaido: Road and Region." *Geographical Review* 27:353–77.
Hanslik, E. 1909. *Biala: Eine Deutsche Stadt in Galizien.* Leipzig: Fischer.
Harris, Britton. 1964. "A Note on the Probability of Interaction at a Distance." *Journal of Regional Science* 5, no. 2: 31–35.
Harris, Chauncey D. 1954. "The Market as a Factor in the Localization of Industry in the United States." *Annals of the Association of American Geographers* 44: 315–48.
Hartshorne, Richard. 1939a. "The Nature of Geography: A Critical Survey of Current Thought in the Light of the Past." *Annals of the Association of American Geographers* 29:173–658.
———. 1939b. *The Nature of Geography.* Lancaster, Pa.: Association of American Geographers.
———. 1959. *Perspective on the Nature of Geography.* Monograph Series, Association of American Geographers No. 1. Chicago: Rand McNally.
Hasse, E. 1891/1892. "Die Intensität grossstädtischer Menschenanhäufungen." *Allgemeines Statisches Archiv* 2, no. 3: 615–30.
Hassert, K. 1907. "Die Städte—geographisch betrachtet." Aus Natur-Und Geisteswelt 163 (Leipzig).
Hassinger, H. 1910a. "Über einige Aufgaben der Geographie der Grossstädte." *Geographischer Jahresbericht Aus Österreich* 8:1–32.
———. 1910b. "Beiträge zur Siedlungs-und Verkehrsgeographie von Wien." *Geographische Gesellschaft Zu Wien, Mitteilungen* 53:5–93.
Hayes, Cordey M., and Wilson, A. G. 1971. "Spatial Interaction." *Socio-Economic Planning Science* 5:73–95.
Hayes, Kingsley; Poston, Dudley; and Schnirring, Paul. 1973. "The Distance and Intervening Opportunities Hypotheses." *Economic Geography* 49, no. 1: 68–73.
Heanue, K. E., and Pyers, C. E. 1967. *A Comparative Evaluation of Trip Distribution Procedures.* Washington, D.C.: Highway Research Board No. 114.
Heine-Geldern, Robert. 1968. "Cultural Diffusion," *International Encyclopedia of the Social Sciences* 4:169–73. New York: Macmillan Co. and Free Press.
Heisenberg, Werner. 1930. *The Physical Principles of the Quantum Theory.* Chicago: University of Chicago Press.
———. 1958. *The Physicist's Conception of Nature.* London: Hutchinson.
———. 1959. *Physics and Philosophy: The Revolution in Modern Science.* London: G. Allen and Unwin.
———. 1961. *On Modern Physics by Werner Heisenberg and Others.* New York: C. N. Potter.
———. 1966. *Introduction to the Unified Field Theory of Elementary Particles.* London: Interscience.
Herbst, Jurgen. 1961. "Social Darwinism and American Geography." *Proceedings of the American Philosophical Society* 105:538–44.
Hettner, Alfred. 1895. "Die Lage der menschlichen Ansiedelungen." *Geographische Zeitschrift* 1:361–75.
———. 1902. "Die wirtschaftlichen Typen der Ansiedelungen." *Geographische Zeitschrift* 8:92–100.
———. 1905. "Das Wesen und die Methoden der Geographie." *Geographische Zeitschrift* 11:545–64, 615–29, 671–86.
Higbee, Edward C. 1952. "The Three Earths of New England." *Geographical Review* 42:425–38.
Hirschman, A. O. 1958. *The Strategy of Economic Development.* New Haven: Yale University Press.
Hudson, John C. 1969. "Diffusion in a Central Place System." *Geographical Analysis* 1:45–58.
Huff, David L. 1959. "Geographical Aspects of Consumer Behavior." *University of Washington Business Review* 18, no. 9: 27–37.
———. 1960. "A Topographical Model of Consumer Space Preferences." *Papers and Proceedings, Regional Science Association* 6:159–74.
———. 1961. "Ecological Characteristics of Consumer Behavior." *Papers and Proceedings, Regional Science Association* 7:19–28.
———. 1962a. *Determination of Intra-urban Retail Trade Areas.* Los Ange-

les: University of California, Graduate School of Business Administration.

———. 1962b. "Note on the Limitation of Intra-urban Gravity Models." *Land Economics* 38:64–66.

———. 1963. "A Probabilistic Analysis of Shopping Center Trade Areas." *Land Economics* 39, no. 1: 81–90.

———. 1964. "Defining and Estimating a Trading Area." *Journal of Marketing* 28:34–38.

———. 1966. "A Programmed Solution for Approximating an Optimum Retail Location." *Land Economics* 42, no. 3: 293–303.

Huntington, Ellsworth. 1915. *Civilization and Climate.* New Haven: Yale University Press.

———. 1924. *The Character of Races.* New York: Charles Scribner's Sons.

———. 1927. *The Human Habitat.* New York: D. von Nostrand Co.

———. 1945. *Mainsprings of Civilization.* New York: John Wiley and Sons.

———, and Carlson, Fred A. 1934. *The Geographic Basis of Society.* New York: Prentice-Hall.

Hyman, G. M. 1969. "The Calibration of Trip Distribution Models." *Environment and Planning* 1:105–12.

Ikle, Fred C. 1954. "Sociological Relationships of Traffic to Population and Distance." *Traffic Quarterly* 8, no. 2: 123–36.

———, and Hammer, Carl. 1957. "Intercity Telephone and Airline Traffic Related to Distance at the 'Propensity of Interact.' " *Sociometry* 20, no. 4: 306–16.

Isaac, Julius. 1947. *Economics of Migration.* London: Kegan Paul, Trench, Trubner and Co.

Isard, Walter. 1956. *Location and Space Economy.* Cambridge, Mass.: M. I. T. Press.

———. 1960. *Methods of Regional Analysis: An Introduction to Regional Science.* Cambridge, Mass.: M.I.T. Press.

———. 1971. "Spatial Interaction: Some Suggestive Thoughts from General Relativity Physics." *Papers and Proceedings, Regional Science Association* 27:17–38.

——— et al. 1972. *Ecologic-Economic Analysis for Regional Development.* New York: Free Press.

Isbell, Eleanor Collins. 1944. "Internal Migration in Sweden and Intervening Opportunities." *American Sociological Review* 9, no. 6: 627–39.

James, Preston E. 1926. "Some Geographic Relations in Trinidad." *Scottish Geographical Magazine* 42:84–93.

———. 1927. "A Geographic Reconnaissance of Trinidad." *Economic Geography* 3:87–109.

———. 1929. "The Blackstone Valley." *Annals of the Association of American Geographers* 19:67–109.

———. 1934a. *An Outline of Geography.* Boston: Ginn.

———. 1934b. "The Terminology of Regional Description." *Annals of the Association of American Geographers* 24:78–92.

———. 1948. "Formulating Objectives of Geographic Research." *Annals of the Association of American Geographers* 38:270–76.

———. 1957. "Changes in the Geography of Trinidad." *Scottish Geographical Magazine* 73:158–66.

———. 1969. "The Significance of Geography in American Education." *Journal of Geography* 68:473–83.

———. 1972. *All Possible Worlds: A History of Geographical Ideas.* Indianapolis: Bobbs-Merrill.

Jarema, F. E. et al. 1968. *Evaluation of Trip Distribution and Calibration Procedures.* Washington, D.C.: Highway Research Board, no. 191.

Jensen-Butler, Christopher. 1972. "Gravity Models as Planning Tools: A Review of the Theoretical and Operational Problems." *Geografiska Annaler* 54B, no. 1: 68–78.

Johans, L. C., and Lindstahl, S. 1968. "Some Models in Marketing Geography with Special Reference to Probability Surfaces." *Acta Geographica* 20, no. 17: 237–53.

Johnston, R. J. 1970. "Latent Migration Potential and the Gravity Models: A New Zealand Study." *Geographical Analysis* 2, no. 4: 387–97.

Jurkat, Ernest H. 1957. "Discussion: Geography of Prices and Spatial Interaction." *Papers and Proceedings, Regional Science Association* 3:134-236.

Katz, Elihu; Levin, Martin L.; and Hamilton, Herbert. 1963. "Traditions of Research in the Diffusion of Innovations." *American Sociological Review* 28:237–52.

Kendall, Henry M. 1935. "The Central Pyrenean Piedmont of France." *Michigan Papers in Geography* 5:377–414.

Kimble, George H. T. 1938. *Geography in the Middle Ages.* London: Methuen and Co.

———. 1951. "The Inadequacy of the Regional Concept." In L. D. Stamp and S. W. Woolridge, eds. *London Essays in Geography.* Cambridge: Harvard University Press, pp. 151–74.

Kirby, Howard R. 1970. "Normalizing Factors of the Gravity Model: An Interpretation." *Transportation Research* 4:37–50.

Kniffen, Fred. 1951. "Geography and the Past." *Journal of Geography* 50:126-29.

Kohl, J. G. 1841. *Der Verkehr Und Die Ansiedelungen Der Menschen In Ihrer Abhangigkeit Von Der Gestaltung Der Erdoberflache.* Leipzig: Arnoldische Buchhandlung.

Kramer, Fritz L. 1967. "Eduard Hahn and the End of the 'Three Stages of Man.' " *Geographical Review* 57:73–89.

Krausse, Gerald H. 1971. "Historic Galena: A Study of Urban Change and Development in a Midwestern Mining Town." *Bulletin of the Illinois Geographical Society.* 13:3–19.

Kuhn, Thomas S. 1971. *The Structure of Scientific Revolutions.* Chicago: University of Chicago Press.

Lakshmanan, T. R., and Hansen, Walter. 1965. "Retail Market Potential Model." *Journal of the American Institute of Planners* 31:134–43.

Langhaar, Henry L. 1951. *Dimensional Analysis and Theory of Models.* New York: Wiley.

Leibnitz, Gottfried W. 1890. *The Philosophical Work of Leibnitz.* New Haven: Tulle, Marehouse and Taylor.

———. 1896. *New Essays Concerning Understanding.* New York: Macmillan Co.

Long, W. H. 1970. "The Economics of Air Travel Gravity Models." *Journal of Regional Science* 10, no. 3: 353–63.

Lovejoy, Arthur O., and Boas, George. 1953. *Primitivism and Related Ideas in Antiquity.* Baltimore: Johns Hopkins University Press.

Luckermann, F., and Porter, P. W. 1960. "Gravity and Potential Models in Economic Geography." *Annals of the Association of American Geographers* 50, no. 4: 493–504.

Maddala, G. S., and Knight, P. T. 1967. "International Diffusion of Technical Change: A Case Study of the Oxygen Steel Making Process." *Economic Journal* 77:531–58.

Mansfield, Edwin. 1968a. *The Economics of Technological Change.* New York: W. W. Norton and Co.

———. 1968b. *Industrial Research and Technological Innovation.* New York: W. W. Norton and Co.

Marble, Duane F. 1959. "Transport Input at Urban Residential Sites." *Papers and Proceedings, Regional Science Association* 5:253–66.

Mason, Joseph B., and Moore, Charles T. 1970. "An Empirical Reappraisal of Behavioristic Assumptions in Trading Area Studies." *Journal of Retailing* 46, no. 4: 31–37.

Mathur, W. K. 1970. "An Economic Derivation of the 'Gravity Law' of Spatial Interaction: A Comment." *Journal of Regional Science* 10, no. 3: 403–5.

Matthews, James S. 1949. "Expressions of Urbanism in the Sequent Occupance of Northwestern Ohio." University of Chicago, Department of Geography Research Paper No. 5.

McCune, Shannon. 1949. "Sequence of Plantation Agriculture in Ceylon." *Economic Geography* 25:226–35.

McGaugh, Maurice E. 1950. "The Settlement of the Saginaw Basin." University of Chicago, Department of Geography Research Paper No. 16.

McVoy, Edgar C. 1940. "Patterns of Diffusion in the United States." *American Sociological Review.* 5:219–27.

Medvedkov, Yu V. 1966. "Concept of Entropy in Settlement Pattern Analysis." *Papers and Proceedings, Regional Science Association* 18:165–68.

Meitzen, A. 1895. *Siedelung Und Agrarwesen Der West-Und Ostgermanen.* Berlin: Hertz.

Merrens, H. Roy. 1965. "Historical Geography and Early American History." *William and Mary Quarterly* 22:529–48.

Meyer, Alfred H. 1936. "The Kankakee 'Marsh' of Northern Indiana and Illinois." *Michigan Papers in Geography* 6:359–96.

———. 1954. "Circulation and Settlement Patterns of the Calumet Region of Northwestern Indiana and Northeastern Illinois (The First Stage of Occupance—The Pottawatamie and the Fur Trader)." *Annals of the Association of American Geographers* 44:245–75.

———. 1956. "Circulation and Settlement Patterns of the Calumet Region of Northwestern Indiana and Northeastern Illinois (The Second Stage of Occupance—Pioneer Settler and Subsistence Economy)." *Annals of the Association of American Geographers* 46:312–56.

Mikesell, Marvin W. 1969. "The Borderlands of Geography as a Social Science." In Muzafer Sherif and Carolyn W. Sherif, eds. *Interdisciplinary Relationships in the Social Sciences.* Chicago: Aldine, pp. 227–48.

Miller, George A. 1947. "Population, Distance and the Circulation of Information." *American Journal of Psychology* 60, no. 2: 276-84.

Minton, Hubert L. 1937. *The Evolution of Conway, Arkansas.* Ph.D. diss., University of Chicago.

Misra, R. P. 1968. *Diffusion of Agricultural Innovations.* Mysore: Prasaranga.

Moore, H. L. 1914. *Economic Cycles: Their Law and Cause.* New York: Macmillan.

Moreno, J. L. 1934. *Who Shall Survive? A New Approach to the Problem of Human Interrelations.* Washington, D.C.: Nervous and Mental Health Publishing Co.

———. 1947. "Contributions of Sociometry to Research Methodology in Sociology." *American Sociological Review* 12, no. 3: 287–92.

———. 1956. *Sociometry and the Science of Man.* New York: Beacon House.

Morrill, Richard L. 1963. "The Distribution of Migration Distances." *Papers and Proceedings, Regional Science Association* 11:75–84.

———. 1968. "Waves of Spatial Diffusion." *Journal of Regional Science* 8:1–18.

Myrdal, Gunnar. 1957. *Economic Theory and Underdeveloped Regions.* London: Duckworth.

Nash, Manning. 1966. *Primitive and Peasant Economic Systems.* Scranton, Pa.: Chandler.

National Economic Development Office. 1970. *Urban Models in Shopping Studies.* London: National Economic Development Office.

Nelson, Howard J. 1949. "The Livelihood Structure of Des Moines, Iowa." University of Chicago, Department of Geography Research Paper No. 4.

Nichols, Vida. 1969. *Growth Poles: An Investigation of Their Potential as a Tool for Regional Economic Development.* Philadelphia: Regional Science Research Institute.

Nicholson, Norman L. 1952. "The Establishment of Settlement Patterns in Ausable Watershed, Ontario." *Geographical Bulletin* 1:1–13.

Niedercorn, J. H., and Bechdolt, V. B., Jr. 1969. "Economic Derivation of the 'Gravity Law' of Spatial Interaction." *Journal of Regional Science* 9, no. 2: 273–82.

———. 1972. "An Economic Derivation of the 'Gravity Law' of Spatial Interaction: A Further Reply and a Reformulation." *Journal of Regional Science* 12, no. 1: 127–36.

Olsson, Gunnar. 1965a. "Distance and Human Interaction—A Migration Study." *Geografiska Annaler* 47B:3–43.

———. 1965b. *Distance and Human Interaction: A Review and Bibliography.* Philadelphia: Regional Science Research Institute.

———. 1967a. "Central Place, Spatial Interaction, and Stochastic Processes." *Papers and Proceedings, Regional Science Association* 18:13–46.

———. 1967b. "Spatial Theory and Human Behavior." *Papers and Proceedings, Regional Science Association* 21:229–42.

———. 1970a. "Explanation, Prediction, and Meaning Variance: An As-

sessment of Distance Interaction Models." *Economic Geography* 46, no. 2 Supplement: 223–33.
———. 1970b. "Logics and Social Engineering." *Geographical Analysis* 2, no. 4: 361–75.
Parkins, Almon E. 1918. *The Historical Geography of Detroit.* East Lansing: Michigan Historical Commission.
Parr, J. B. 1965. *The Nature and Function of Growth Poles in Economic Development.* Seattle: University of Washington.
Passarge, S., ed. 1930. *Stadtlandchaften Der Erde.* Hamburg: de Gruyter.
Pedersen, Poul Ove. 1970. "Innovation Diffusion Within and Between National Urban Systems." *Geographical Analysis* 2:203–54.
Pemberton, H. Earl. 1936. "The Curve of Culture Diffusion Rate." *American Sociological Review* 1:547–56.
Perloff, Harvey S. 1960. *Regions, Resources and Economic Growth.* Baltimore: Johns Hopkins University Press.
Perroux, François. 1955. "Note sur la Notion de Pôle de Croissance." *Economie Appliquée* 8:307–20.
Pfouts, Ralph W. 1958. "Discussion: Population Projection by Means of Income Potential Models." *Papers and Proceedings, Regional Science Association* 4:155–58.
Poincaré, Henri. 1892–1899. *Les Méthodes Nouvelles De La Mécanique Céleste.* (Paris: Gouthier et Fils.
———. 1905a. *Lecons de Mécanique Céleste Professées à la Sorbonne.* Paris: Gauthier Villars.
———. 1905b. *Science and Hypothesis.* London: Walter Scott Co.
———. 1907. *The Value of Science.* New York: Science Press.
———. 1914. *Science and Method.* Translated by Francis Marland. New York: T. Nelson and Sons.
Polanyi, Karl. 1944. *The Great Transformation: The Political and Economic Origins of Our Time.* New York: Rinehart.
———. 1957. "The Economy as Instituted Process." In Karl Polanyi, Conrad M. Arensberg, and Harry W. Pearson, eds. *Trade and Markets in the Early Empires: Economics in History and Theory.* Chicago: Free Press, pp. 243–70.
Pounds, Norman J. G. 1952. *The Ruhr: A Study in Historical and Economic Geography.* Bloomington: Indiana University Press.
Pred, Allan. 1966. *The Spatial Dynamics of U. S. Urban-Industrial Growth, 1800–1914.* Cambridge, Mass.: M.I.T. Press.
Prince, Hugh C. 1971. "Real, Imagined and Abstract Worlds of the Past." *Progress in Geography No. 3.* London: Edward Arnold, pp. 4–86.
Pyle, Gerald F. 1969. "The Diffusion of Cholera in the United States in the Nineteenth Century." *Geographical Analysis* 1:59–75.
Quandt, R. E. 1965. "Some Perspectives of Gravity Models." *Studies in Travel Demand.* Washington, D.C.: Department of Commerce.

Ratzel, Friedrich. 1882. *Anthropogeographie 1: Grundzüge Der Anwendung Der Geographie Auf Die Geschichte.* Stuttgart: J. Engelhorn.
———. 1891. *Anthropogeographie 2: Die Verbreitung Des Menschen.* Stuttgart: J. Engelhorn.
———. 1898. *The History of Mankind,* vol. 3. London: Macmillan and Co.
———. 1903. "Die geographische Lage der grossen Städte." *Jahrbuch Der Gehe-Stiftung,* pp. 31–72.
———. 1921. *Anthropogeographie.* 2 vols. Stuttgart: J. Engelhorn.
Radcliffe-Brown, A. R. 1957. *A Natural Science of Society.* Glencoe, Ill.: Free Press.
Ravenstein, Ernest G. 1885. "The Laws of Migration." *Journal of the Royal Statistical Society* 48:167–235.
———. 1889. "The Laws of Migration." *Journal of the Royal Statistical Society* 52:241–305.
Redlich, Fritz. 1953. "Ideas—Their Migration in Space Over Time." *Kyklos* 6:301–22.
Reilly, William J. 1929. *Methods for the Study of Retail Relationships.* University of Texas Bulletin No. 2944. Austin: University of Texas.
———. 1931. *The Law of Retail Gravitation.* New York: W. J. Reiley Co.
Reynolds, Robert B. 1953. "A Test of the Law of Retail Gravitation." *Journal of Marketing* 17, no. 3: 273–77.

Rich, John L. 1920. "Cultural Features and the Physiographic Cycle." *Geographical Review* 10:297–308.

Richards, Wilfred G. 1948. "The Settlement of the Miami Valley of Southwestern Ohio." Ph.D. diss., University of Chicago.

Richthofen, Frh. von. 1908. *Vorlesungen Urber Die Allegemeine Siedlungs—Und Verkehrsgeographie.* Berlin: Reimer. Revised and edited by O. Schlüter.

Ritter, C. 1833. "Über das historische Element in der geographischen Wissenschaft." Reprinted in *Abhandlungen zur Bergründung Einer Mehr Wissenschaftlichen Behandelung Der Erdkunde,* 1852.

———. 1862. "Begriff der Erdkunde als Wissenschaft." In H. A. Daniel, ed. *Allgemeine Erdkunde.* Berlin: Reimer.

Rogers, Everett M. 1962. *Diffusion of Innovations.* New York: Free Press.

Rostow, W. W. 1960. *The Stages of Economic Growth.* Cambridge: Cambridge University Press.

Sack, Robert D. 1973. "A Concept of Physical Space in Geography." *Geographical Analysis* 5, no. 1: 16–34.

Sahlins, Marshall D. 1965a. "On the Sociology of Primitive Exchange." In Michael Banton, ed., *The Relevance of Models for Social Anthropology* (London: Tavistock), pp. 139–236.

———. 1965b. "Exchange-Value and the Diplomacy of Primitive Trade," in June Helm, ed. *Essays in Economic Anthropology, Dedicated to the Memory of Karl Polanyi.* Proceedings of the 1965 Annual Spring Meeting of the American Ethnological Society. Seattle: University of Washington Press, pp. 95–129.

———. 1972. *Stone Age Economics.* Chicago: Aldine-Atherton.

Sauer, Carl O. 1915. *The Geography of the Ozark Highland of Missouri.* Chicago: Geographic Society of Chicago.

———. 1925. "The Morphology of Landscape." *University of California Publications in Geography* 2:19–53.

———. 1927. "Anthropogeography." In E. C. Hayes, ed. *Recent Developments in the Social Sciences.* Philadelphia: Lippincott.

———. 1952. *Agricultural Origins and Dispersals.* New York: American Geographical Society.

———. 1963. *Land and Life.* Berkeley: University of California Press.

Scharf, J. H., ed. 1970. "Struktur und Funktion." *Nova Acta Leopoldina, Abhandlungen, N. F.* 194, no. 35: 341–90.

Schlüter, O. 1899. "Bemerkungen zur Siedelungsgeographie." *Geographische Zeitschrift* 5:65–84.

———. 1906a. *Die Ziele der Gegraphie Des Menschen.* Munchen: Bruckmann.

———. 1906b. "Die leitenden Geschichtspunkte der Anthropogeographie, insbesondere der Lehre F. Ratzel." *Archiv der Sozialwissenschaften Und Sozialpolitik* 22.

Schneider, Morton. 1959. "Gravity Models and Trip Distribution Theory." *Papers and Proceedings, Regional Science Association* 5:51–56.

Schockel, Bernard H. 1947. "Manufactural Evansville, 1820–1933." Ph.D. diss., University of Chicago.

Semple, Ellen Churchill. 1903. *American History and Its Geographical Conditions.* Boston: Houghton, Mifflin and Co.

———. 1911. *Influences of Geographic Environment.* New York: Henry Holt and Co.

Seneca, Joseph, and Cicchetti, Charles J. 1969. "A Gravity Model Analysis of the Demand for Public Communications." *Journal of Regional Science* 9, no. 3: 459–70.

Servan-Schreiber, Jean Jacques. 1969. *The American Challenge.* Harmondsworth, Middx.: Penguin Books.

Shrode, Ida M. 1948. "The Sequent Occupance of the Rancho Azusa De Duarte: A Segment of the Upper San Gabriel Valley of California." Ph.D. diss., University of Chicago.

Siebert, Horst. 1969. *Regional Economic Growth: Theory and Policy.* Scranton, Pa.: International Textbook Co.

Smith, J. Russell. 1935. "Are We Free to Coin New Terms?" *Annals of the Association of American Geographers* 25:17–22.

Soja, Edward W. 1968. *The Geography of Modernization in Kenya.* Syracuse, N.Y.: Syracuse University Press.

Sorensen, Clarence W. 1951. "The Internal Structure of the Springfield, Illinois, Urbanized Area." University of Chicago, Department of Geography Research Paper No. 20.

Sorokin, Pitrim A. 1928. *Contemporary Sociological Theories Through the First Quarter of the Twentieth Century.* New York: Harper and Row.

Stanislawski, D. 1946. "The Origin and Spread of the Grid-Pattern Town." *Geographical Review* 36:105–20.

Steward, Julian H. 1955. *Theory of Culture Change: The Methodology of Multilinear Evolution.* Urbana: University of Illinois Press.

Stewart, Charles T. 1960. "Migration as a Function of Population and Distance." *American Sociological Review* 25, no. 3: 347–56.

Stewart, John Q. 1941. "An Inverse Distance Variation for Certain Social Influences." *Science* 93:89–90.

———. 1947. "Empirical Mathematical Rules Concerning the Distribution and Equilibrium of Population." *Geographical Review* 37, no. 3: 461–85.

———. 1948a. "Concerning Social Physics." *Scientific American* 178, no. 5: 20–23.

———. 1948b. "Demographic Gravitation: Evidence and Application." *Sociometry* 2, nos. 1–2: 31–58.

———. 1950. "The Development of Social Physics." *American Journal of Physics* 18, no. 5: 239–53.

———. 1952. "A Basis for Social Physics." *Impact of Science on Society* 3, no. 2: 110–33.

———. 1958. "Discussion: Population Projection by Means of Income Potential Models." *Papers and Proceedings, Regional Science Association* 4:153-54.

———, and Warntz, William. 1958a. "Macrogeography and Social Science." *Geographical Review* 48, no. 2: 167–84.

———. 1958b. "Physics of Population Distribution." *Journal of Regional Science* 1, no. 1: 99–123.

Stouffer, Samuel A. 1940. "Intervening Opportunities: Theory of Mobility and Distance." *American Sociological Review* 5:845–67.

———. 1960. "Intervening Opportunities and Competing Migrants." *Journal of Regional Science* 2, no. 1: 1–26.

Sweet, Frank H. 1964. "An Error Parameter for the Reilly-Converse Law of Retail Gravitation." *Journal of Regional Science* 5, no. 2: 69–72.

Tatham, G. 1951. "Environmentalism and Possibilism." In G. Taylor, ed. *Geography in the Twentieth Century.* New York: Philosophical Library.

Taylor, Griffith. 1945. "The Seven Ages of Towns." *Economic Geography* 21:156–60.

Thoman, Richard S. 1953. "The Changing Occupance Pattern of the Tri-State Area: Missouri, Kansas, Oklahoma." University of Chicago, Department of Geography Research Paper No. 31.

Thomas, Lewis F. 1931. "The Sequence of Areal Occupancy in a Section of St. Louis, Missouri." *Annals of the Association of American Geographers* 21:75–90.

Thompson, Donald L. 1963. "Subjective Distance." *Journal of Retailing* 39, no. 1: 1–6.

Thompson, Wilbur. 1968. "Internal and External Factors in the Development of Urban Economics." In H. S. Perloff and L. Wingo, eds. *Issues in Urban Economics.* Baltimore: Johns Hopkins University Press for Resources for the Future, pp. 43–62.

Thünen, J. H. von. 1826. *Der Isoliate Staat in Beziehung Auf Landwirtschaft Und Nationalökonomie.* Hamburg: Olbnicht. Translated into English by C. M. Wartenberg. 1966. *Isolated State.* New York: Pergamon. Edited and with an introduction by P. Hall.

Trewartha, Glenn T. 1940. "A Second Epoch of Destructive Occupance in the Driftless Hill Lands." *Annals of the Association of American Geographers* 30:109–42.

Troll, C. 1947. "Die geographische Wissenschaft in Deutschland." *Erdkunde* 1:3–48.

Tuan, Yi-Fu. 1968. *The Hydrologic Cycle and the Wisdom of God.* Toronto: University of Toronto Press.

Turner, Frederick Jackson. 1894. "The Significance of the Frontier in American History." *Annual Report of the American Historical Association for 1893,* pp. 199–227.

———. 1920. *The Frontier in American History.* New York: Henry Holt and Co.

Uhlig, H., ed. 1972. *Die Siedlungen Des Ländlichen Raumes.* Giessen: Lenz Verlag. (= Materialien zut Terminologie der Agrarlands-chaft 2—English, French, German.)

Ullman, Edward L. 1943. "Mobile: Industrial Seaport and Trade Center." Ph.D. diss., University of Chicago.

———. 1957. *American Commodity Flows.* Seattle: University of Washington Press.

University of Glasgow Social and Economic Studies. 1968. *Regional Policies in Efta: An Examination of the Growth Center Idea.* Edinburgh: Oliver and Boyd.

Valkenburg, Samuel van. 1944. *Elements of Political Geography.* New York: Prentice-Hall.

Vining, Rutledge. 1949. "The Region as an Economic Entity and Certain Variations to be Observed in the Study of Systems of Regions." *American Economic Review* 39, no. 3: 89–104.

Voorhees, Alan M. 1957. "Discussion: Geography of Prices and Spatial Interaction." *Papers and Proceedings, Regional Science Association* 3:130–33.

Wagner, Philip L. 1960. The *Human Use of the Earth.* Chicago: Free Press. Press.

———, and Mikesell, Marvin W. 1962. *Readings in Cultural Geography.* Chicago: University of Chicago Press.

Wagon, D. J., and Hawkins, A. F. 1970. "The Calibration of the Distribution Model for the SELNEC Study." *Transportation Research* 4:103–13.

Warntz, William. 1956. "Measuring Spatial Association with Specific Consideration of the Case of Market Orientation of Production." *Journal of the American Statistical Association* 51:597–604.

———. 1957a. "Contributions Toward a Macroeconomic Geography: A Review." *Geographical Review* 47, no. 3: 420–24.

———. 1957b. "Geography of Prices and Spatial Interaction." *Papers and Proceedings, Regional Science Association* 3:118–29.

———. 1959a. "Geography at Mid-Twentieth Century." *World Politics* 2:442–54.

———. 1959b. "Progress in Economic Geography." In P. E. James, ed. *New Viewpoints in Geography.* Washington, D.C.: National Council for the Social Studies, pp. 54–75.

———. 1967. "Global Science and the Tyranny of Space." *Papers and Proceedings, Regional Science Association* 19:7–22.

Weber, A. 1909. *Über den Standort der Industrien 1: Reine Theorie Des Standorts.* Tübingen: Mohr.

Wells, H. G. 1902. *Anticipations of the Reaction of Mechanical and Scientific Progress Upon Human Life and Thought.* New York: Harper and Brothers.

Wheeler, Jesse H., Jr. 1950. "Land Use in Greenbrier County, West Virginia." University of Chicago, Department of Geography Research Paper No. 15.

White, C. Langdon. 1965. "Sequent Occupance in the Santa Clara Valley, California." *Journal of the Graduate Research Center, Southern Methodist University* 34:277-99.

Whittlesey, D. 1929. "Sequent Occupance." *Annals of the Association of American Geographers* 19:162–65.

———. 1931. "The Urbanization of a Farm Village." *Annals of the Association of American Geographers* 21:142–43 (abstract).

———. 1933. "Coastland and Interior Mountain Valley: A Geographical Study of Two Typical Localities in Northern New England." In John K. Wright, ed. *New England's Prospect.* New York: American Geographical Society.

———. 1945. "The Horizon of Geography." *Annals of the Association of American Geographers* 35:1–36.

———. 1956. "Southern Rhodesia: An African Compage." *Annals of the Association of American Geographers* 46:1–97.

Wilson, A. G. 1967. "A Statistical Theory of Spatial Distribution Models." *Transportation Research* 1:253–69.

———. 1968. "Models in Urban Planning: A Synoptic Review of Recent Literature." *Urban Studies* 5, no. 5: 249–76.

———. 1969a. "Notes on Some Concepts in Social Physics." *Papers and Proceedings, Regional Science Association* 22:159–94.

———. 1969b. "The Use of Entropy Maximizing Models in the Theory of Trip Distribution, Mode Split and Route Split." *Journal of Transport Economics and Policy* 3, no. 1: 108–26.

———. 1970a. "Inter-Regional Commodity Flows: Entropy Maximizing Approaches." *Geographical Analysis* 2, no. 3: 255–82.

———. 1970b. "The Use of the Concept of Entropy in System Modeling." *Operational Research Quarterly* 21, no. 2: 247–65.

———. 1971. "A Family of Spatial Interaction Models, and Associated Developments." *Environment and Planning* 3:1–32.

Wright, Alfred J. 1936. "The Industrial Geography of the Middle Miami Valley, Ohio." *Michigan Papers in Geography* 6:401–27.

Wright, John Kirtland. 1925. *The Geographical Lore of the Time of the Crusades.* New York: American Geographical Society.

———. 1944. "Human Nature in Science." *Science* 100:299–305. Reprinted in John K. Wright. 1966. *Human Nature in Geography.* Cambridge: Harvard University Press, pp. 53–67.

———. 1962. "Miss Semple's 'Influences of Geographic Environment': Notes Toward a Bibliography." *Geographical Review* 52:346–61. Reprinted in John K. Wright. 1966. *Human Nature in Geography.* Cambridge: Harvard University Press, pp. 188–204.

Wrigley, Robert L., Jr. 1942. "The Occupational Structure of Pocatello, Idaho." Ph.D. diss., University of Chicago.

Zipf, George Kingsley. 1941. *National Unity and Disunity: The Nation as a Bio-Social Organism.* Bloomington, Ill.: Principia Press.

———. 1942. "Unity of Nature, Least Action and Natural Social Social Science." *Sociometry* 5:48–62.

———. 1946a. "Some Determinants of the Circulation of Information." *American Journal of Psychology* 59, no. 3: 401–21.

———. 1946b. "The P_1P_2/D Hypothesis: On the Intercity Movement of Persons." *American Sociological Review* 11, no 5: 677-86.

———. 1946c. "The P_1P_2/D Hypothesis: The Case of Railway Express." *Journal of Psychology* 22:3-8.

———. 1949. *Human Behavior and the Principle of Least Effort: An Introduction to Human Ecology.* Cambridge, Mass.: Addison-Wesley.

/G70.N37>C1/